FREEZING
FISH AT SEA
A HISTORY

By J J Waterman

Edinburgh: Her Majesty's Stationery Office

HMSO publications are available from:

HMSO Bookshops

13a Castle Street, Edinburgh, EH2 3AR   (031) 225 6333
49 High Holborn, London, WC1V 6HB   (01) 211 5656
(Counter service only)
258 Broad Street, Birmingham, B1 2HE   (021) 643 3740
Southey House, 33 Wine Street, Bristol, BS1 2BQ   (0272) 24306/24307
9–21 Princess Street, Manchester, M60 8AS   (061) 834 7201
80 Chichester Street, Belfast, BT1 4JY   (0232) 238451

HMSO Publications Centre
(Mail and telephone orders only)
PO Box 276, London, SW8 5DT
Telephone orders (01) 622 3316
General enquiries (01) 211 5656

HMSO's Accredited Agents
(see Yellow Pages)

And through good booksellers

ISBN 0 11 492485 6

CONTENTS

Man has been aware for centuries of the preservative effect of low temperatures, and millions of words have been written about the application of cold to the keeping of foods. The history of refrigeration in general is well documented, and there are a number of detailed accounts of the development of refrigeration technology in the food industry. This essay is an attempt to bring together the stories of those events that are specifically related to the freezing of fish, and to the transfer of that concept to the confines of a ship.

Methods of chilling fish at sea, whether by means of ice or mechanical refrigeration, are deliberately excluded, as are methods designed to induce partial freezing by reducing the temperature of fish a little below that of melting ice, although some of the early attempts at freezing described here were in practice little more than what would today be called superchilling. Having said that, a few of the more important events in the history of mechanical refrigeration and the manufacture of ice are briefly referred to as precursors of ways of achieving the lower temperatures needed for the long term preservation of fish as food.

It was necessary, when selecting what to include in this account, to decide what is meant here by the freezing of fish 'at sea'. Strictly speaking, a factory ship anchored in a loch on the west coast of Scotland or in a sheltered Alaskan harbour can hardly be said to be at sea; nevertheless 'at sea' in the context of this essay means on board ship, whether or not the ship herself catches fish, provided she is located near to the point of capture of the fish she is processing. Floating factories remote from the fishing grounds, and thus providing a service no different from any landward installation, are considered not to be freezing fish at sea.

The main emphasis throughout is on the fisheries of the north Atlantic as prosecuted from western Europe, with a strong bias towards UK interests, but relevant events in North America are included. Developments in the freezing at sea of shrimp, crab, tuna and Pacific salmon are largely excluded, since the fisheries, and markets, for those species demanded solutions to problems that were markedly different from those encountered in the white fish and herring industries of the north Atlantic.

Events in this story are divided roughly into three periods. From the middle of the nineteenth century to the end of the First World War there evolved the art of mechanical refrigeration, and with it the first thoughts of applying it to the preservation of fish at sea. From 1919 to the end of the Second World War was a period of experimentation in which freezing plants were tried on almost every kind of ship from sailing boats to secondhand passenger liners, mostly without success. After 1945, with vastly improved knowledge of the art of the possible, ships were equipped with installations that worked and, after a great deal of dogged development work, actually made a profit.

The commissioning of a new seagoing venture is usually an occasion for much enthusiasm and rosy publicity; such events are fully recorded in the contemporary technical press. The demise of an unsuccessful venture is less newsworthy; ships are laid up, perhaps for years awaiting disposal, or freezing plant is removed and a ship quietly returns to more conventional activities; operators rarely want to advertise the reasons for experimental or commercial failure. Thus this narrative is incomplete; all too often all that can be said is that a particular ship was apparently unsuccessful. For how long her owners persevered in the face of increasing financial loss, or for what technical, economic or social reasons they had to admit defeat may never be known. But occasionally in this story, which is largely about who did what and when, there are opinions expressed, by those who were in a position to know or guess, about how and why.

Palaeolithic man probably kept his food in a cool cave, and the Egyptians and the Chinese made and used natural ice three thousand years ago, but it was not until 1755 that William Cullen first demonstrated at Glasgow University the controlled production of cold by evaporation of a fluid. Already the word 'first' has been used here to define an historical event, a perpetual trap for the unwary recorder. Any listing of the firsts of Watt and Stephenson for example in the history of the steam engine has to be set about with definitive qualification before the claims can stand close scrutiny, and the history of refrigeration science and technology is no different; there are many firsts, in thermometry, calorimetry, thermodynamics and engineering, already well catalogued elsewhere. The choice of date at the head of this section therefore, 1834, is a somewhat arbitrary one. It signifies the invention of what was subsequently referred to as 'the parent of all modern compression machines', and is meant to serve only as a marker from which to start the story.

Jacob Perkins

Perkins (1766-1849) was an American who is best remembered in his home country as the inventor of a process for steel engraving of banknotes. He came to England in 1819 with a party of engravers to compete for a Bank of England prize for a forgery-proof system of making banknotes, and stayed to experiment in steam engines and the liquefaction of gases; by 1822 he claimed to have been successful in liquefying air. In later years he became interested in refrigeration and in 1834 applied for a patent for an 'Improvement in the Apparatus and Means of Producing Ice and Cooling Liquids'. His design, for a closed circuit in which ethyl ether was compressed and condensed, was converted into a freezing machine of practical form by collaboration with John Hogue, a reputable London engineer, and together they demonstrated it in 1835. Perkins, like so many others, was before his time; his invention had no immediate appeal except as an interesting scientific toy.

Freezing mixtures: Benjamin and Grafton

Although it was to be another 20 years before Perkins' condensing unit became a commercial product, other ways were known of achieving low temperatures. The literature of nineteenth century patent applications is littered with ideas for using freezing mixtures, but credit for what must be among the earliest of recorded inventions for the freezing of fish must go to Henry Benjamin of St Mary-at-Hill, a London fish factor, and his associate Henry Grafton of Chancery Lane. Their patent application was 'Inrolled in the Inrolment Office, July, 1842' and was described as follows:

> This invention consists in preserving animal and vegetable matters by the application of freezing mixtures, for the purpose of freezing or cooling them.

The patentees commence their specification by describing the method pursued by them for preserving fish, which is as follows: 'The fish is placed in a copper or other metallic vessel, which is filled with cold water, and deposited in a wooden trough, containing a mixture of one part salt, and six parts of pulverized ice. When the fish has become frozen, it may be preserved for a considerable time, by continuing the application of the freezing mixture and thus keeping it in a frozen state. The patentees state . . . that, although they prefer the freezing mixture to be composed of salt and ice, yet other freezing mixtures may be used.'

Fish in a cold climate

Fish frozen naturally in the cold winters of North America and Russia had for long been commodities of commerce, and from the 1840s, as land and sea transport rapidly improved, reached the larger towns as cheap but not very desirable food. Freezing was sometimes adventitious but, on the coasts of Canada and Newfoundland for example, nature was encouraged by laying fish out in thin layers on beaches or on wooden platforms exposed to the cold wind, and turning the fish at intervals. There are less well substantiated accounts of the natural freezing processes being extended to fishing vessels; certainly the line fishermen of the north west Atlantic were accustomed to freezing some of their catch for subsequent use as bait, and occasionally a ship would arrive in port with frozen fish aboard. Some of the frozen catch of the *Flying Cloud* of Gloucester, Massachusetts, was reported as having been hawked successfully in the streets of Boston in 1855. There are similarly vague accounts of Gloucester schooner captains landing frozen fish in 1856 and 1857, but whether freezing was entirely natural, or was assisted, perhaps by the use of salt and ice, is not known.

More salt and ice: Piper and Pike

James Harrison, a Scotsman who had emigrated to Australia, and of whom more later, was reported to be experimenting in 1855 with ice and salt mixtures for freezing meat and fish, having successfully made ice by mechanical refrigeration. In the USA, the first patent application of note for a fish freezing process was taken out by Enoch Piper of Camden, Maine, in 1861. He claimed a more practical method of applying a mixture of ice and salt. His trade was mainly in salmon. The fish were laid on a wooden tray in the bottom of an insulated box, leak-proof containers of ice and salt were placed over the top of the fish, and the box closed. The mixture was renewed after 12 hours, and the fish were considered to be frozen after 24 hours. He advocated glazing after freezing, by dipping the fish in cold water, and storage in a cold store also refrigerated by ice and salt. His cold store design was the subject of a second patent in 1862. Although he was reluctant to allow others in the fish trade to become licensees of his patents, the ice and salt technique, typically with the fish inside lidded metal pans, had by 1865 become common commercial practice at some US ports.

There quickly followed the idea of extending the method by using it on board ship, and in 1866 Charles Pike of Providence, Rhode Island, took out a patent to this effect, although there is no record of what practical success was enjoyed.

Advances in refrigeration: Carré and Tellier

Among the many who experimented with the mechanical production of cold in the middle of the nineteenth century, two French engineers made significant contributions. Ferdinand Carré (1824-1900) made the first successful absorption refrigeration machine; having patented his design in 1859, he built a machine that worked in 1861. He thus provided an optional source of artificial cold, an option that was taken up by some of those engaged in the 1870s in the battle for supremacy in marine refrigeration.

Charles Tellier (1828-1913) made a compressor that used methyl ether as a refrigerant, and in 1868 applied his machinery in the chocolate, brewing and meat transport industries. He studied the effect of cold on a range of foods, and is reported to have been one of the first to attempt the application of mechanical refrigeration to fish, although he advocated chilling and the maintenance of a dry surrounding atmosphere rather than freezing for both meat and fish. His equipment was used on the *Frigorifique* to carry meat, described later.

Frozen fish: a Victorian consumer's view

Such frozen fish products as had reached the market by the 1860s had not generated much enthusiasm; frozen slowly and stored at temperatures not much below 0°C, they compared unfavourably with the best of fresh fish. One contemporary view was expressed in

Wholesome Food: or the Doctor and the Cook by E S and E J Delamere, published in London in 1868:

> '. . . frozen fish is neither good for soup nor anything else. The northern nations eat it out of necessity, not choice. The convenience of transport gained is more than outbalanced by the deterioration of quality. The effects of frost are to rend the tissues, and to reduce them to a tasteless pulp, whose very wholesomeness is questionable. If such is the case, with the most substantial meats, it is still more so with delicate muscle, like that of fish. Meat or fish, once frozen, soon putrefy after thawing; they should be used immediately or not at all. Refrozen, it is difficult to ascertain their actual condition, and their consumption may cause serious derangement to health.'

The ice trade

As artificial refrigeration equipment gradually became available in the 1860s and early 1870s, most of the efforts to make use of it were concentrated on chilling and, in particular, the manufacture of ice. In areas where natural ice was not available, as in the southern states of the USA, icemaking had become an important commercial enterprise by the 1870s, but early attempts to supply the North Sea fishing industry from plants at Barking and Shadwell were unsuccessful, largely due to continued competition from natural ice, mainly imported from Norway at prices well below those of the manmade product. Further improvements in the efficiency and safety of mechanical refrigeration were needed before artificial chilling, let alone freezing, could become the norm.

Harrison and the meat trade

The potential profit in the carriage of meat from Australasia and South America to Europe was enormous, and engineers who were anxious to exploit the possibilities of marine refrigeration concentrated almost exclusively for many years on the problems of shipping chilled and frozen meat. Development work throughout the 1870s was successful, and resulted in the rapid growth of a number of refrigeration companies, expert in marine work, who subsequently turned their attentions to other commodities, including fish.

James Harrison (1816-1893) was the son of a Scottish salmon fisher who emigrated to Australia in 1837. Apparently unaware of the earlier work of Perkins, he patented an ethyl ether compressor in 1855, and arranged for its manufacture both in London and in Sydney. His machine and Carré's absorption machine were probably the only two practical models available anywhere in the world in the 1860s, first for icemaking and subsequently for freezing. Harrison, anxious to expand the Australian meat trade, at first experimented with freezing mixtures, adding salt to the ice he could now readily produce, but his first trial shipment by this method to the UK on the sailing ship *Norfolk* in 1873 was spoilt by a brine leak. Tellier, who had tried his methyl ether machine unsuccessfully on a meat ship in 1868, was more successful when he tried it again on a voyage of the *Frigorifique* from Rouen to Buenos Aires and back in 1876-77, although the temperature of carriage was only just below freezing point; weight loss from the meat in dry air was high. Carré's absorption machines were used in the *Paraguay* on a round voyage from Marseille to Buenos Aires and back to Le Havre in 1877-78, and in 1879 the *Circassia* carried beef from the USA to London with the aid of a Bell-Coleman cold air compressor. The first successful meat run from Australia was by the *Strathleven* in 1879-80, and from New Zealand by the *Dunedin* in 1882; both ships carried cold air machines. Thus the carriage of frozen food by sea became established practice, and the new international trade, managed largely from the UK, resulted in the rapid growth of cold storage facilities at principal ports, served by fleets of refrigerated cargo vessels. Meat had the advantage over fish, of course, in being a more robust commodity in the frozen state but, more importantly, the carrier had only to keep cold what had already been frozen; meat could be frozen after slaughter in an adjacent landbased plant before shipment, whereas fish after slaughter had first to be frozen on the high seas before cold storage and carriage home.

Carl von Linde's first successful ammonia compression refrigerating machine

Progress in marine refrigeration was hampered in the early 1870s by the lack of safe and efficient refrigerants for use in vapour compression systems; ether was a frightening prospect aboard ship. Although the use of carbon dioxide had been demonstrated in 1866, there had been no commercial exploitation, and it was the ammonia machine that first became established.

Carl von Linde (1842-1934) was a German professor of engineering who, having studied the work of Perkins, Carré and others, published in 1871 his *Improved design of an Ice Production and Refrigerating Engine*. Egged on by the brewers of Munich, he built a machine to his design in 1875, still using ether, but in 1876 he changed to ammonia, and an improved machine in 1877 was so promising that he began commercial production of it in 1878. He also licensed other manufacturers throughout Europe and North America, and by the turn of the century Linde had made over 4000 ammonia compressors for use in cold stores, ice factories, breweries and ships.

St. Clement and other trawlers

There was still considerable ignorance in the 1880s about freezing in relation to fish, and a number of unsubstantiated reports of that time about 'freezing' at sea were really references to chilling or perhaps some degree of partial freezing as a means of slightly extending shelf life. Sporadic attempts were still being made to use mixtures of ice and salt at sea, both in Europe and in the USA, but what may have been the first mechanically refrigerated and insulated fishroom in a trawler was reported to have been installed in the steam trawler *St. Clement* when she was built by Hall Russell in Aberdeen in 1883 for Thomas Walker, a local trawler owner; a cold air compressor was used. This was almost certainly a chilling plant.

By 1886 a number of steam trawlers working out of Hull and Grimsby were reputed to have been fitted with cold air machines and insulated fishrooms, as had possibly one or more of the steam carriers being used to transport fish from the North Sea fleets to Billingsgate, but none of these could have been anything more than an auxiliary chilling plant. Doubtless they were ineffective even in that role, and would soon have been removed or allowed to fall into disuse, as indeed were their twentieth century counterparts installed in large numbers of UK trawlers in the 1950s.

J & E Hall: the Dartford influence

The firm of J & E Hall was founded at Dartford Ironworks in Kent in 1785, and became renowned for a wide range of mechanical engineering work, but particularly for steam boilers and beam engines. When a cold air machine designed by Giffard was exhibited in Paris in 1877, J & E Hall became interested in the possibilities of refrigeration engineering, and a Giffard machine was brought to Dartford in 1878. For a few years they made cold air machines of an improved type for marine and other purposes, but in 1888 one of their senior engineers, Marcet, developed a compressor that worked successfully on carbon dioxide as a refrigerant, and this rapidly became their standard machine for shipboard use; the extremely inefficient cold air machine immediately became obsolescent.

Many of the subsequent advances in the technology of freezing food were in part attributable to that small select band of companies who, having become expert in producing sources of cold, enthusiastically set about the task of creating new applications in this novel engineering field. J & E Hall designed brine freezing equipment to complement their steam-driven carbon dioxide compressor, and in 1889 Everard Hesketh, Hall's company chairman, and Alexander Marcet took out a patent for equipment to freeze foods, including fish, by direct immersion in refrigerated brine. Another UK company took out a patent in the same year in the names of Douglas and Donald for a

Giffard's cold air machine

Cold air machine built by J & E Hall,
c 1885

The Dartford Works of J & E Hall
circa 1885
The works were established a hundred years previously by John Hall in 1785

process of freezing fish by placing them in bags surrounded by water in
a can, around which was to be circulated mechanically refrigerated
brine. This idea of an 'ice can' freezer was revived by Friedrichs in 1915
and by Petersen in 1922. Neither of the freezing methods patented in
1889 was taken up commercially at the time.

Henri Rouart was a French refrigeration engineer from the pioneer
company of Mignon & Rouart, who made both absorption and com-

pression machinery. He too sought to encourage increased application of refrigeration, and in 1898 took out a British patent for a brine immersion freezing process for fish; his method differed from that of the 1889 Hesketh & Marcet patent in that he proposed to use a solution of glycerol, alcohol and salt instead of sodium chloride brine as the cold liquid in which the fish were to be immersed.

Sharp freezing

By 1890 the Norwegians had a fish freezing plant in Bergen that employed the ice and salt mixture, and several other plants were built in Scandinavia that relied on the long established freezing mixture technique, but in North America there was a revival of interest in the oldest, simplest, and perhaps the slowest method of freezing, once mechanical refrigeration became more readily available. All that was required was to lay the fish out in a cold place, and wait. Insulated rooms were built, with internal piping coiled horizontally to form shelves on which trays or lidded pans of fish could be laid; any fish too big for this treatment were simply laid on the floor. The shelf grids were cooled either by pumping cold brine through them or allowing refrigerant vapour to expand in them, and the fish were slowly frozen, to a slight extent by direct contact between grid and tray but largely by natural convection of cold air; fan assistance was exceptional. This process, which became known as sharp freezing, was used in a number of North American plants, one of the first being that at Sandusky, Ohio, in 1892. Although this was in essence slow freezing inside a cold store—freezing times were typically 12 hours or more due to poor heat transfer—the method prevailed and, indeed, still exists. It was the forerunner of the air blast freezer and, in its hybrid form of cold shelf and cold air, reappeared in more sophisticated versions, such as the Fairfreezer of 1946, but the sharp freezer in its elementary form was too slow and too wasteful of space ever to be a serious contender for shipboard freezing practice. Furthermore the product all too often displayed the worst features of frozen fish, freezer burn, denaturation, rancidity, and severe loss of weight through surface evaporation.

Having said all that, by 1894 the first shipments of Pacific salmon, often frozen by this method, were finding their way from British Columbia to the UK and the trade grew rapidly. The continued importation of frozen salmon and halibut from the Pacific was subsequently one of the factors that motivated British operators to compete in the 1920s with factory mother ships in the north west Atlantic.

Brine freezing takes off: Ottesen and others

In spite of the efforts of J & E Hall, Mignon & Rouart, and other makers of refrigeration plant, nobody in the fish trade seemed to want their brine freezers. There were doubts about product acceptance, a great lack of understanding about what went on during the process, and as always the fear of risking one's capital in a pioneering process and losing it.

J R Henderson of London believed that the sudden immersion of fish in cold brine was detrimental to the product, and in 1910 he patented a brine freezing process that included a precooling period in cold air before immersion, but much the same idea had been the subject of an earlier patent by T D Kyle in 1905. Henderson also tried patenting his process in the USA in 1913, but brine freezing found little favour at that time in the land of the sharp freezer.

It fell to Anton Jensenius Andreas Ottesen (1860-1936) to make some kind of breakthrough. He was a fish exporter from Thisted, Denmark, who was convinced that freezing in ice-salt mixtures or in saturated brine resulted in too great an uptake of salt in the fish, and he proposed immersion in unsaturated brine that was reduced in temperature almost to its freezing point. He applied for his first Danish patent in 1911, conducted further experiments in 1912 and was granted his patent in 1913. In the event, comparative tests conducted under practical conditions both in the UK and in the USA failed to substantiate his claims that salt uptake in the product was markedly less, but more importantly his publicity was good. He promoted his method enough to persuade the Danish refrigeration company Sabroe to take up manufacture, and within a decade there were a dozen land-based plants with brine tanks to his design in Europe and the USA. Fish frozen by the Ottesen process, or by any brine immersion process for that matter, were quite soon considered to be superior to the products of sharp freezing or of ice-salt mixtures, largely because of the rapidity of the process. The shortcomings of brine frozen fish were still in evidence, however, principally poor appearance and discoloration after long term storage; knowledge of appropriate storage times and temperatures for any kind of frozen fish was at this time still largely nonexistent.

Ottesen was certainly a successful pioneer of quick freezing, successful in that he got beyond the stages of an idea and a patent to make the innovatory step into commercial practice. By 1915 he even had his plant on board ship.

Brine freezing: variations on a theme

In Norway an attempt was made in 1912 to improve the performance of the ice and salt routine; Nekolai Dahl, a fish merchant in Trondheim, allowed the brine formed by the freezing mixture to trickle down through open crates of fish to give improved contact between refrigerant and product. By 1913 he had a commercial plant on this system, and later there was another on the west coast of the USA. Dahl also proposed the use of his method in ships' holds, and some years afterwards a namesake of his did put a modified system on board ship.

Some critics considered Dahl's treatment, later known as the Frigus system, to be no more than a useful chilling expedient in localities that did not justify investment in refrigeration machinery. Certainly it was far less effective than complete immersion; indeed some thought it to be even slower than the ice-and-salt pan-freezing method it was meant to replace.

Henrik Jansen Bull, director of the Fisheries Research Laboratory at Bergen, took out a number of patents from 1913 onwards for a freezing method akin to the ice can technique for making block ice, after the manner of Douglas and Donald back in 1889, but with ice and salt instead of mechanical refrigeration as the source of cold. Metal moulds containing fish were set in a tank through which was pumped brine from a mixture of ice and salt. Bull experimented with a variety of moulds to improve heat transfer, but the cold source remained the same.

Bull's method apparently failed to find commercial acceptance, but in 1915 Martin Friedrichs of Hamburg used such a technique specially for freezing eels. He employed a deep narrow tapered can of water in which the fish were suspended on a frame, and immersed the can in a bath of ice-and-salt brine.

More than 30 years later the mould or ice can containing fish and water was the subject of extensive experiment before the present day form of the vertical plate freezer evolved from it.

Karmøy

Perhaps this ship deserves to be much better known, because she was probably the first fishing vessel ever to be equipped to freeze her catch at sea. In 1915 the steam fishing vessel *Karmøy* of Haugesund went to Esbjerg to be fitted by Sabroe with a carbon dioxide compressor and an Ottesen brine tank installation, claimed to be capable of handling 10 tons of fish a day. Little is now known about her performance, except that the venture was unsuccessful, and the plant was subsequently removed.

Freezing theory: Plank

This account of the first attempts to freeze fish ends on an optimistic note. Hitherto the coldmakers and the traders in fish had sought to find, mostly by trial and error, ways and means of prolonging shelf life by reducing the temperature of fish to below that of melting ice, and to some extent had succeeded. But virtually nothing was known about the effects that freezing and cold storage had upon the appearance and eating quality of the commodity, or the reasons why the reputation of frozen fish was no higher than it had been in 1868, when it had been declared a tasteless pulp to be eaten only out of necessity, not choice.

Rudolf Plank (1886-1973) and his successors changed all that. Plank, who began teaching refrigeration in Danzig in 1913, conducted systematic scientific work in Germany on the freezing of foods from 1915 onwards. In 1916 Plank, Ehrenbaum and Reuter published the results of their first study of the brine freezing process, which demonstrated the importance of speed of freezing in relation to texture and quality. Designers of freezing equipment, already aware of the obvious need for speed to achieve higher output, now had the added incentive of higher quality. The term quick freezing entered the language.

EXPERIMENTS AT SEA 1918~1945

Ten years were to elapse after the *Karmøy* experiment of 1915 before fish were frozen at sea again, but 10 years is not a surprisingly long time in that most conservative sphere of commerce, the fish industry. Development of the freezing process continued on shore but, as fishing trips became longer and quality at first sale on the quayside became poorer, it was increasingly obvious that belated application of refrigeration at the port would all too often mean preservation of the second rate. Every now and then, as the landward cold chain was gradually built up, designers and developers would look longingly to sea, and would occasionally persuade a shipowner either to invest in, or permit the demonstration of, another shipboard venture. Some 30 odd ships, of disparate size and purpose, were tried between the Wars; most are described in what follows.

Brine freezing in the UK: Hardy and Piqué

For almost 30 years after the J & E Hall patent in 1889 there had been virtually no British initiative in the freezing and cold storage of fish. Icemaking flourished at the larger fishing ports; the demand from trawlers making ever longer voyages was proving insatiable. Cold stores sprang up at the principal cargo ports to house imported meat, and even Billingsgate had rudimentary storage for imports of frozen fish from North America, but nowhere in the UK did anyone freeze fish.

The government-financed Food Investigation Board, under its research director, William Hardy, recognized the need for work on fish preservation, and in 1918 paid for the establishment of experimental brine freezing plant in the premises of J M Tabor, a Billingsgate fish merchant and importer. Hardy had one engineer at his disposal, a Belgian named Jean Julien Piqué, and he put him to work with a colleague, Adair, on the Billingsgate freezer, mainly to find a solution to the problem of saving fish, principally herring, in times of glut. The experiments continued for the best part of two years until the total absence of daylight in their workplace, the damp and the noise made them both seriously ill. In 1920 the small and inadequate freezer had to be dismantled and removed to make way for reconstruction of Billingsgate basement, in readiness for the brand new cold stores and ice plant of the London Ice & Cold Storage Co. Opened in 1921, it offered storage at $-10°C$, and could produce 50 tons of ice a day at about £3 per ton.

Piqué and his freezer moved to Cambridge, where the work began again in 1922 at the recently completed Low Temperature Research Station, but active work soon dwindled at a site so far removed from a fishing port, and further experimentation had to wait until Hardy's dream of a coastal research station came nearer to reality. Piqué meanwhile turned his attention to the drawing board, and produced a design for a trawlerborne brine freezer that was patented on behalf of the government by the Imperial Trust, Hardy & Piqué in 1922. Piqué proposed to tip the fish into an open mesh drum revolving in a tank of sodium chloride brine, refrigerated by a steam driven ammonia compressor.

He advocated washing and glazing after freezing, before storage at −20°C. The advantages claimed were low first cost, quick freezing through good contact, economy of labour for handling the catch, and no weight loss during freezing. The disadvantages were misshapen fish giving a poor stowage rate, salt uptake by the product, poor appearance after prolonged storage, and corrosion of moving parts immersed in brine.

Piqué's brine drum freezer for herring

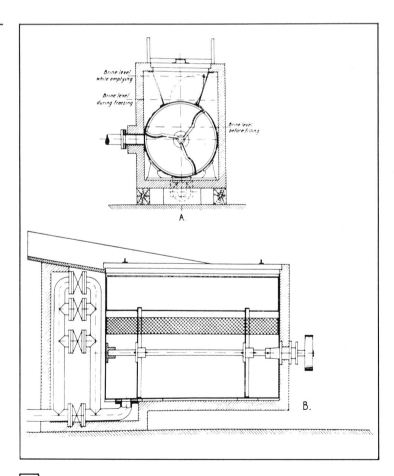

20

Piqué's design was ignored completely by the British fishing industry. There is some evidence to suggest that a plant to his design was installed experimentally in a ship working off the coast of India, probably at government expense, but it was left to the French, who obtained a licence to exploit the patent, to put the system on a commercial vessel.

More brine freezers: Mann and others

Piqué's proposals for trawlerborne equipment were preceded by a 1920 patent taken out by Robert Mann. He proposed a deep upright tank with a hatch opening on the foredeck of a trawler, through which the fish would be lowered, in baskets inside a metal cage, into mechanically refrigerated and filtered brine. The method, which was not obviously different from Ottesen's and other brine tank systems, apparently remained a paper study.

There were also designs for continuous brine freezers at this time; that of E de Goër de Hervé was probably the first in 1920. He proposed carrying fish through a long brine tank on a continuous wire mesh belt fitted with paddles. The Compagnie d'Alimentation & d'Installations Frigorifiques saw fit to patent it, and there it would seem the matter rested.

An astute Estonian called Zarotschenzeff took out his first British patent in 1921 for the 'Z' process of brine spray freezing, but a fuller account of the application of his method in ships must wait until a little later in this story.

Advances in the USA

Paul W Petersen, a Chicago refrigeration engineer, having studied the designs of Ottesen and Piqué, proposed a switch from direct to indirect brine freezing so that brines with lower freezing points than sodium chloride could be used to achieve faster freezing. Ludwig Hirsch had advocated the use of calcium or magnesium chloride brine in Germany in 1921 but, since he had suggested spraying directly onto fish, not surprisingly his proposals had been ignored.

Petersen clung to the by then almost standard icemaking equipment, and simply changed the refrigerant in the tank to calcium chloride. He filled tapered metal cans with fish, apparently without additional water, immersed them in the tank, and after freezing dipped the cans in water and inverted them to release the blocks as in conventional icemaking practice. He claimed there were no expansion problems as the fish froze, and he had the system working in the Bay City Freezer Company's factory in 1922. Although Petersen never had shipborne equipment in mind, here again was a crude forerunner of the vertical plate freezer of the 1940s. His main concern was to improve upon the notoriously slow performance of US sharp freezers, and to this end he succeeded.

Harden F Taylor, at one time chief technologist in the US Bureau of Fisheries and subsequently a senior manager of Atlantic Coast Fisheries, patented a brine freezer in 1923. He proposed a continuous tunnel freezer in which brine sprays played on the fish as they passed through. Although an experimental model was built in a Washington laboratory, it was considered to be too complicated and expensive to be commercially viable when handling the cheaper species of fish. Perhaps his best contribution to progress was his comprehensive review of progress in freezing for the US Commissioner of Fisheries in 1926, which was published under the title *Refrigeration of Fish* in 1927.

By the time that world experts exchanged views at the Fourth International Congress of Refrigeration in June 1924 there were commercially available, apart from the notorious sharp freezer, three kinds of freezing plant, all utilizing brine as the refrigerant—the tank in which fish were frozen individually by direct immersion, as represented by the Ottesen plant, the tank in which fish were frozen in blocks in moulds, as represented by the Petersen plant, and the brine trickle process in which fish were frozen in what might be described as lumps, as represented by the Dahl or Frigus system, still used in Scandinavia for herring. But Birdseye changed all that.

Consumer products: Birdseye and Cooke

Further modest improvements continued to be made on brine freezers for shore operation; for example R E Kolbe, son of a fish merchant in Erie, Pennsylvania, devised means of freezing in calcium chloride brine in 1925 by using lidded pans of fish in which entrapped air under the lids kept the brine away from the fish. But Clarence Birdseye (1886-1956), who took out his first freezer patent in 1924, swung away from brine immersion to pioneer the plate freezer. Although the work of this New York inventor was aimed at freezing consumer products rather than the raw material, it eventually had some effect on the pattern of development of shipboard freezing equipment; some mention of his achievements is therefore appropriate here. His first freezer was an adaptation of the ice can; since he was freezing fillets it had to be an indirect method. He used a small can that did not require a mechnical hoist and fitted it with an internal frame to form a block of fillets. About 1927 he changed to a double belt of monel metal, refrigerated by calcium chloride brine, and passed his frames of fillets between the belts, freezing being achieved by direct contact between cold metal and the flat faces of the fillet blocks. From there he went on to make the multiple plate 'froster', as it was called, with its double walled refrigerated plates in tiers, the still completely recognizable forerunner of the horizontal plate freezer as it is known today; it was patented by Birdseye and Hall in 1929, the year that the Birdseye Seafoods Company was taken over by Postum, later to become General Foods; that company in turn marketed and publicized the portable multiplate froster in its commercial form.

A H Cooke of Atlantic Coast Fisheries, New York, also tackled the freezing of fillets, and in 1926 devised a system of jacketed or double walled pans, inside which were stacked trays of fillets in frames. Calcium chloride refrigerant was pumped round the jackets, and hot brine was used when freezing was completed to defrost and release the frames. Aiming at the same consumer outlets as Birdseye, he also made a variant of the horizontal plate freezer. A battery of hollow cast aluminium shelves were fitted inside an insulated box, and trays of fillets were laid on each shelf. When calcium chloride was pumped through the shelves, the fillets were cooled both by conduction through the tray and by convection of cold air. In this form it represented the intermediate stage of development between the primitive sharp freezers with their pipe grid shelves and the Fairfreezer of 1946.

Salmon and tuna

Fish supplies destined for Cannery Row had quality specifications markedly different from fish entering the conventional distribution chain. Much of the Pacific catch of salmon and tuna was heat processed in cans, and the principal fishing grounds were a long way from Seattle or Vancouver, north to Alaska for salmon, and south to the tropics for tuna. Refrigeration at sea was obviously highly desirable, and keeping times longer than were possible at chill temperature were needed to get the fish into the cannery cold stores in reasonable condition. Salt uptake was relatively unimportant—salt would be added anyway during the canning process—and, since the fish were to be chopped up and cooked, appearance of the whole fish prior to processing was also of less importance. Thus brine immersion freezing for what were mostly large fish proved attractive.

By about 1925 some of the schooners that acted as carriers from the tuna boats working in the Gulf of Mexico had brine immersion tanks on board, and later the catchers themselves carried tanks. Although the process was described as freezing, storage temperature was relatively high, freezing times for such large fish were long, and for many years the technique was little more than superchilling. In a typical installation there was a series of tanks or wells containing sodium chloride brine that was cooled by pipe grids round the tank walls. Each tank was filled with fish from a deck hatch, and once the fish were considered to be frozen the brine was pumped out and transferred to another tank; the fish were then kept dry, using the well as a cold store. The Japanese too had mother ships or carriers serving their wide ranging tuna lining fleets, and by 1924 some of these had brine freezers aboard.

Brine freezing was first tried in the Alaskan salmon fishery about 1926. The Seattle Siberian Fish and Cold Store Co owned the motor factory ship *Apollo*, which had been used for carrying salted salmon. It was equipped with an Ottesen brine tank freezer capable of handling about ½ ton an hour; the salmon were lowered in baskets through hatches in the foredeck, and residence time in the brine was only 1-2 hours. The frozen fish were wrapped in parchment and packing paper before being put in wooden cases in the ship's cold store. The product was said to be for the European market. In the spring of 1926 it was announced that 'the expedition will start from Kamtchatka in collaboration with the Russian Government and a Japanese company.' The expedition's fate is not known.

Janot

The French rights to Piqué's 1922 patent were acquired by the Société Anonyme Francaise pour la Conservation des Poissons par la Congélation, variously known for short as SAFPC or SAFCP. By the autumn of 1925 SAFPC had prepared plans for a fleeting operation off the Spanish and Moroccan coasts to supply frozen sardines to the 200-ton cold store the company had in Spain.

They used the *Janot*, 46 metres in length over all, as a mother ship, and converted her by the summer of 1926. She was equipped with two cold

The French steamer *Janot*, equipped with a Piqué drum freezer in 1926

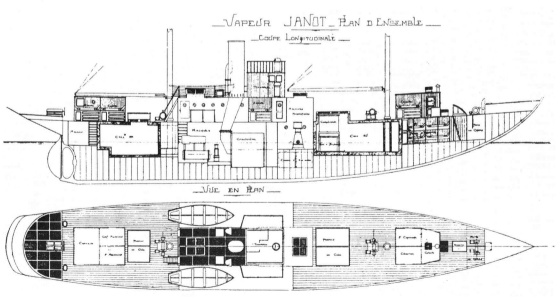

stores refrigerated by carbon dioxide compressors and capable of holding a total of 100 tons of frozen fish at $-10°C$. Freezing was brine immersion, using a single drum freezer to Piqué's design; its capacity was ½ ton of fish an hour using brine at $-18°C$, cooled by an ammonia compressor. Although the *Janot* was described as a converted steam trawler, it was not intended that she herself should fish in this operation.

> 'This ship can accompany the trawlers and collect their catch. From this fact the trawlers will leave without ice, with a stock of . . . fuel which can exceed . . . their usual supply; the radius of operations is thus considerably increased. The ship will be of service chiefly in countries where fish is abundant and the means of transport insufficient. It could thus collect the sardines on the coasts of Morocco, Portugal and Spain by receiving it on deck, and thus saving small boats the obligation of carrying it to a distant port . . . it will allow fishing in large quantities.'

The vision thus conjured up of hordes of steam trawlers making long trips in hot pursuit of the humble sardine is almost certainly an erroneous one; inadequate contemporary translations in the technical press are probably to blame. Nevertheless the *Janot* did actually receive and freeze some fish, somewhere, but the product was declared to be of unacceptably poor appearance, and the venture was abandoned.

Helder

Just about the time the French were equipping *Janot*, the UK trade press was hinting at a 'mystery ship' being fitted out in Hull; the first *Helder* expedition was being prepared.

The ss *Helder* began life as the Dutch passenger ship *Madura*, built in Amsterdam in 1897, and subsequently became a Norwegian refrigerated meat carrier. She was 300 feet long, with a gross tonnage variously quoted as 3450 and 3378 tons; other undefined tonnage figures were bandied about in the press, ranging from 2000 to 5000. Hellyer Brothers, prominent trawler owners in Hull, had numerous consultations in early 1926 with William Hardy's Food Investigation staff and with the refrigeration company of J & E Hall about the prospects of brine freezing fish at sea. They were sufficiently encouraged to buy the *Helder* from Norway and bring her to Hull for fitting out, and to borrow J J Piqué from Hardy to act as technical adviser. The Norwegians almost certainly had a financial stake in the venture; indeed one account describes it as being a Norwegian expedition, although this may be explained by the fact that Norwegians did most of the fishing.

The main purpose of the *Helder* was to receive and freeze line-caught halibut from dories working off Greenland. To this end she was fitted with J & E Hall carbon dioxide compressors to run an Ottesen brine tank installation that could handle up to 1¼ tons of fish an hour. Her cold stores, probably those previously installed for meat, had a capacity of 60 000 cubic feet, and were probably operated at about $-10°C$. She could carry up to 240 people on board, most of them Norwegian fishermen to man the 48 small dories she carried, with 4-5 men to a boat. Working in the open sea in depths down to 400 metres, the dories could not work in bad weather, and transfer of catch to the mother ship was not possible in winds much above force 3. Steam liners also worked to the mother ship, and assisted dorymen with line hauling when the weather worsened.

The *Helder* left Hull for Greenland on her first voyage in June 1926 and returned to Hull's Albert Dock in October with 650 tons of frozen halibut and cod. She probably collected fishermen from Norway on the outward passage and dropped them off again on the way home, as she did on most subsequent voyages. The *Helder* then lay in dock throughout the winter, dispensing daily amounts of frozen fish, quaintly known as quantums, at prices fixed slightly below those of the fresh fish market.

The *Helder* worked well throughout the summer voyage of 1927, some of her frozen product being sent home by carrier before she herself returned, and the last of her catch was still being landed and sold in Hull in March 1928. Steam liners and carriers working to the mother ship at Greenland were reported as occasionally replenishing her fuel supply, but just how this was achieved is not clear; probably *Helder* moved into sheltered waters or even into port to provide what must have been badly needed social breaks in the long summer voyage. Piqué paid a visit himself to the fleet at about this time, at the request of Hellyer's, going out and back apparently on steam liners.

Helder made a third voyage which ended with her arrival in Albert Dock on 29 September 1928, the first season in which she had been accompanied by her sister ship, described later. Her movements from then on are less clear; she was by then less of a news story, and she certainly had some maintenance and crewing problems from time to time. She probably made the 1929 voyage, the year in which she was renamed *Arctic Prince*, but was certainly laid up in King George V Dock in Hull from July 1930 until April 1932, when she sailed to Aalesund to pick up her dorymen en route to Greenland once more. On the 1933 voyage there was a sudden switch in crewing arrangements; she went to Newfoundland in April to pick up 200 local fishermen there in place of the customary Norwegians. Her final summer voyage was in 1935, arriving back in Hull with frozen halibut in October, and in the spring of 1936 *Arctic Prince* left for Murmansk, having been sold to Russia to serve as a floating cold store, where she joined her already retired sister ship *Arctic Queen*. Some possible reasons for the rise and fall of the Hellyer venture are given after the section on the *Arctic Queen*.

Oscar Dahl and the *Pen Men*

After the conversion of the *Janot* the next French project was reported on early in 1927. The Société Maritime de Pêche Industrielle at Le Havre was said to be equipping a ship called the *Calgary* with brine freezing plant for fish and shellfish, together with fish meal and oil plant, possibly for working with the French salting fleet off Newfoundland and St Pierre, but no further details are available.

Down in La Rochelle, however, a trawler owner and entrepreneur named Oscar Dahl revived the idea of his Norwegian namesake, Nekolai Dahl, first patented in 1913. Oscar Dahl, in collaboration with E Kjorstad, another Norwegian, modified the original brine trickle system for shipboard use. Known as the Kjorstad-Dahl method, K-D or 'Kadyfreez', the fish were packed in latticework cases over which ice-salt brine was pumped at about $-18°C$; the brine trickled down among the fish and, after a 'freezing' time of 2-3 hours, the fish were held in the fishroom at -3 to $-4°C$ until discharge at the port of landing. Dahl called them 'frosted fish' and, as is described later, defended the efficacy of his system in comparison with other shipboard treatments when it was alleged that his was no more than an extension of chilling.

Dahl put the K-D system on one of his own trawlers, the *Pen Men*, and put the catch on display when the ship arrived at La Rochelle in November 1927. The *Pen Men* continued to demonstrate for a few months, until in the spring of 1928 she sank in La Rochelle harbour. Although she was subsequently raised, the K-D plant was destroyed.

Ben Meidie

There was a ripple of interest in freezing at sea among British trawler owners in the mid 1920s. Captain Lawford of the Iago Steam Trawling Co for example, looked at the possibilities of its use on the long southerly voyages from Milford Haven in search of hake, and sought advice from William Hardy and his staff in 1926; Crampin of Grimsby, many of whose ships made long trips as steam liners, asked refrigeration firms for proposals. Partly as a result of these enquiries Hardy initiated a series of experiments in the summer of 1928, which were managed for him from Aberdeen by Adrian Lumley in cooperation with a committee of trawler owners.

Two commercial steam trawlers were chartered in April 1928 by Hardy's parent government body, the Department of Scientific and Industrial Research (DSIR). There were two aims, to improve current shipboard chilling practice, and to make preliminary investigations of freezing at sea. The ships were the *Cicely Blanche*, which had just been bought by Lawford and renamed *Peter Carey* on transfer to Milford Haven, and the *Ben Meidie* owned by Irvin in Aberdeen. The former need figure no further in this account, for it seems certain that she never carried freezing plant; her work was confined to icing experiments from Milford.

The *Ben Meidie*, hired at a cost of £100 a month plus owner's out-of-pocket expenses, made a number of landings between June and September, proceeds from sale of fish going to DSIR. Apart from conducting icing experiments for direct comparison with *Peter Carey*, she also carried a small freezing plant to Piqué's design and landed sample lots of brine frozen fish for inspection by the trade. Brine at about −20°C, cooled by an ammonia compressor, was used to freeze cod, haddock and hake and hold them in a ship's cold store at about the same temperature; on shore they were transferred to commercial cold storage at −10°C for up to three months. Lumley and his colleagues wrote afterwards that although the fish were good to eat after about one month in store, they then rapidly took on a poor appearance, becoming brown in colour, and were totally unsuitable for smoking, since they failed to take on the customary gloss. High storage temperature was thought to be largely to blame, and it was decided that much more research work on shore was necessary to establish optimum freezing and storage conditions before further seagoing work was attempted. This work was to be done at Torry Research Station, opened in 1929 at Aberdeen under Lumley's direction.

Arctic Queen

Hellyer Brothers of Hull were sufficiently happy with the performance of the *Helder* after two summer voyages to buy another and larger passenger ship for conversion in 1928. Built in 1909 at Middlesbrough as the *Vasari*, and subsequently named *Vauban*, she had worked to South America, first for Lamport & Holt and latterly for the Liverpool, Brazil and River Plate Steam Navigation Co. She was 400 feet long, and 10 078 gross tons. She was equipped with brine tanks of the Ottesen pattern, together with washing and glazing tanks, and carried her frozen catches, mainly halibut, in cold stores running at just below −20°C, which could hold up to 4000 tons. She carried 50 five-man dories with her, and was reported to have about 440 men on board for her first voyage. That first trip was late in starting and she was away only a month or so before returning to Hull at the end of September 1928.

Thereafter she sailed late April or early May and returned in September each year, releasing her frozen cargo at the rate of about 30 tons a day throughout the winter months; no shore cold storage was

Mother ship *Arctic Queen*, operated by
Hellyer's from 1928 to 1935

used. She apparently made a voyage each summer up to and including
1934, picking up and dropping off her Norwegian crew each year, in
much the same manner as *Helder*, latterly *Arctic Prince*. In 1931 she
was reported as having had a particularly poor season when she paid off
the Norwegians in Bergen at the beginning of September 1931. In July
1935 she was sold to Russia, and was joined the following year in
Murmansk by *Arctic Prince*.

Among reasons given at the time for ending the mother ship experiment were: depletion of halibut stocks, high operating costs in terms
of fuel and labour, the difficulties of running and maintaining a complex
factory operation at sea, the problems of transshipment in bad
weather, high capital cost, and a comparatively limited market for large
amounts of frozen halibut at an economic price. Many of these
problems have bedevilled all subsequent high seas fleeting operations
with factory mother ships but transshipment has been the main
stumbling block; in general, either the factory ship has to be capable of
catching its own supplies, or transfer has to be from a fishery that is
close to sheltered waters, as with the current mackerel transfers in
Loch Broom and Carrick Roads. Transfer is markedly different from
that in whaling, as Lochridge of *Fairfree* and *Fairtry* fame was later to
point out; whales are large, they float, and what is more they make
remarkably good fenders between catcher and mother ship in bad
weather.

Sacip

In 1929 the French were sufficiently successful with demonstration
freezing equipment on a conventional steam trawler that what might be

called a production run of four trawlers followed at intervals up until the start of the Second World War.

The Société Anonyme pour la Conservation Industrielle du Poisson (SACIP), took a standard Castle class steam trawler, 30 metres long, and fitted her with a carbon dioxide compressor, a brine freezer that could handle about 0.4 tonnes an hour at $-20°C$, and a cold store to hold 40 tonnes at $-15°C$. The freezer consisted of two compartmented drums like bucket wheels, with a hollow axle through which brine was pumped to the compartments. Fish were loaded into an upper segment, and the drum was turned to bring the next segment to the loading position. Segments were immersed in brine in the lower positions, and speed of rotation was such that freezing was completed during one revolution, which took from ½ to 2½ hours depending on contents. The equipment bore only a faint resemblance to the pioneer Piqué drum freezer as used on *Janot*.

The *Sacip*, or *Sacip 1* as she was apparently later named, was working on grounds much favoured by the German fleet, Viking Bank in the northern North Sea, in the first half of 1929, and during this period made at least one demonstration landing at a German fishing port. By 1930 the *Sacip* was reported to be working from Boulogne under the auspices of the Compagnie Anonyme Francaise de Pêche et d'Armement (CAFPA), who were the concessionaires of the SACIP process, and CAFPA were already planning equipment for her successor, the *Jean Hamonet*.

Blue Peter

Over on the western Atlantic seaboard, at about the same time as Hellyer's bought their second mother ship, a Newfoundland company, Job Brothers of St John's, in whom the Hudson Bay Company had an interest, bought a large secondhand ship for conversion in 1928. Built in 1899 for the River Plate meat trade, and named *Highland Laird*, she was rechristened *Blue Peter* and fitted out to work as a depot ship along the Canadian coast.

The *Blue Peter*, 4132 gross registered tons, and with a storage capacity of 2600 tons, was equipped with ammonia compressors, an Ottesen freezer to handle 25 tons of fish in 24 hours at $-19°C$, and cold storage running at $-10°C$. She was expected to work mainly on salmon and halibut, which the Hudson Bay Company would ship frozen to the UK market. Some canning facilities were also provided, and 150 extra hands were carried for factory work. Supplies were expected to be brought to her by an accompanying fleet of small steamers and fishing vessels, some of the steamers in turn acting as carriers from inshore fishermen.

She was in operation by the summer of 1929, and there were soon complaints that some of the fish supplies reached her several days after capture, and often held without ice during transit from inshore boat to small steamer to factory ship. The operators were also unhappy about direct immersion freezing, claiming that the brine quickly became dirty, and that the frozen product was often of poor appearance. She froze 1000 tons of salmon in 1929, but by 1931 her output of fish was reported to have fallen considerably, she having been used to freeze blueberries in season; she probably worked almost entirely at anchor in sheltered waters and, since collection of fish appears to have been from a wide catchment area, it is doubtful if she met the definition of freezing 'at sea' as given in the introduction.

Zarotschenzeff and the Z process

M T Zarotschenzeff, from Reval, Estonia, had been engaged in refrigeration work on transport as early as 1913 in Russia; he remained active in this field when he went to the USA in 1918, and by 1921 had developed and patented a system of freezing in a 'fog of atomized brine'; in 1926 the Z process as it became known was used in the USA for freezing meat, and by 1928 a demonstration plant had been put into the fish processing plant of H Smethurst & Co in Grimsby. Claimed to be capable of handling 6-12 tons of fillets a day, it was still in use in 1931. But Zarotschenzeff believed his system was particularly applicable to shipboard freezing, and in early 1929 he installed a plant on a salting vessel owned by a French salt cod company at Fécamp, in time for it to make the summer cod fishing voyage to Greenland and Newfoundland. The *Zazpiakbat* was equipped with a small tunnel in which fish were laid on trays or hung on hooks, depending on their size, and sprayed with brine at −21°C through special nozzles to produce a fine mist. Capacity was reported as being 1-3 tonnes a day. The ship left Fécamp in April 1929 and returned to La Rochelle in September with about 10 tonnes of cod and a few halibut stored at −15°C; the main bulk of the catch had been salted in the traditional manner. One of the uses of freezing at sea as envisaged by the French at this time was an ancillary means of preservation on salting voyages to the north west Atlantic.

The Italians also equipped a fishing vessel with a Z plant in 1929, the 42-metre *Naiade* which fished in the Mediterranean off the North African coast. An ammonia compressor supplied the brine heat exchanger and the cooling coils of cold stores, and brine at −16°C was pumped via the atomizers on to the fish, which were either in latticed cases on shelves or, if large, hung on hooks. Although there were four cold stores meant to hold a total of 50 tonnes, apparently only small sample lots were frozen for inspection ashore after storage at about −7°C.

The French, reasonably satisfied by the *Zazpiakbat* demonstration, were prepared to try further. The Société d'Études pour l'Amélioration de la Pêche Maritime, known as SAP for short—the French penchant for unmanageable titles made acronyms almost compulsory—fitted out another Fécamp trawler belonging to the Compagnie de la Morue Francaise, the *Gure Herria*, sometimes referred to by her earlier name of *Mulhouse*, with a brine spray tunnel. She made the summer voyage to Greenland in 1930 and returned with 100 tonnes of frozen fish in her 150-tonne cold store, in addition to her catch of salt fish. A third French salter, the *Izarra*, is believed to have been equipped in 1931, and soon after, probably in 1933, the largest salting trawler of her day, the *Marcella*, built in 1929 and 70 metres long over all, was equipped by the Z licensees, SAP, with a plant to handle what could not readily be salted, especially high value large halibut. The plant was similar to its forerunners, with an ammonia compressor, atomised brine at $-18°C$, and a capacity of about 4 tonnes of fish a day.

The Norwegians too tried a shipboard Z plant, this time for freezing what was described as 'fillets' of halibut, probably steaks or pieces, during a salting voyage by the *Lesseps* from Aalesund to Greenland in the summer of 1931. The product was to be wrapped and boxed on board, for subsequent retail sale in the UK, it being allowed to thaw during inland distribution. Freezing capacity was 1-1½ tons a day.

Sterilex and Nunthorpe Hall

Apart from the small *Ben Meidie* experiment at the cost of the tax-payer, the true British trawler had so far remained uninvaded by freezers, and the average trawler owner remained totally unmoved by the blandishments of Hardy, Lumley and others, as witness the blunt opinions of Marsden, quoted below. Sterilex was a Manchester based company trying to exploit a brine freezing system patented by R A Cowtan, a Manchester stockbroker, and their demonstrations ashore and afloat did little in the event to further the cause among trawlermen.

A brine freezing plant on the Sterilex system was first demonstrated at Mudd's fish processing factory in Grimsby in the summer of 1929, but the patentees remained secretive about exactly what in their treatment was different from other brine freezing methods; their publicity material merely advertised brine freezing without going into detail, but there was apparently a sterilizing wash before the freezing began.

The *Nunthorpe Hall*, a Fleetwood steam trawler that eventually carried a Sterilex freezer, remains almost as mysterious as the method. At the end of 1928 she was reported to have been sold to South Africa, and was to sail for Cape Town, but over a year later she was still at Fleetwood, and during 1929 had been fitted with Sterilex plant which in January 1930 was tried out on a six-day fishing trip to the Minch. She was described as having four compartments called ovens, in which the fish after gutting were laid on shelves and treated with sterilized water; the crew themselves were said to be working in sterilized air! There followed a period of about 20 minutes in 'brine water'—for the fish, not the crew—before freezing proper began at a rate of about 1 ton an hour. Some years later Adrian Lumley, when addressing audiences about the dangers of inadequate freezing, used to quote some unspecified awful example of a British trawler where the catch had been congealed in a solid mass; the story may have been apocryphal, but with few British freezer trawlers to choose from at that time, it may well have been the quaint treatment on the *Nunthorpe Hall* that he referred to. After further alterations following the Minch fishing trials, the *Nunthorpe Hall*, still reported as being Fleetwood-owned, sailed in May 1930 for the Canaries; she was to land refrigerated catches at Las Palmas for shipment by refrigerated carrier to a Continental port, much as the Japanese were doing in Las Palmas 30 years later. It seems unlikely that she ever reached South Africa to fish for hake; after 2 years the plant was abandoned as being too costly in time and manpower, and the trawler reverted to conventional iced stowage.

A Grimsby trawler owner's views

It may be pertinent at this juncture to interpose the views of a prominent UK trawler owner on the prospect of freezing fish at sea. Sir John Denton Marsden, who was in 1929 the head of Consolidated Fisheries, one of the largest trawling companies in the world, had been asked by Sir William Hardy to read a proof copy of the report prepared by Lumley and his colleagues on the 1928 *Ben Meidie* trials, and in his comments Marsden said:

> 'I have always regarded with apprehension the efforts . . . to introduce a satisfactory means of preserving fish by . . . refrigeration; for it seems to me that if any of these efforts should become successful, a tremendous blow would be struck at the Steam Trawling Industry of this country', apparently by the entry of competitive imports from all parts of the world, as already foreshadowed by Canadian frozen salmon and South African hake; no thought of British owners indulging in the hateful method themselves seems to have entered his mind. He went on to say 'I hope it will never come about that the British public will be taught to eat and appreciate frozen fish.' He raised one specific objection to adoption of the method on trawlers: '. . . a plant (to freeze at sea) having the capacity to deal with the first haul before the second is deposited on deck would be so large that no existing trawler would have sufficient space for stowing . . . the fish.'

This fear, which persisted in the views expressed by trawler owners in the UK right up to the 1960s, of being confronted by the largest haul ever rumoured to have been caught, and having to match it with a freezer of commensurate capacity, was never justified; first, claims about the size of such a haul were usually exaggerated, after the tradition of all good fishermen, and secondly the arguments in support of acceptable delays in chill storage of white fish awaiting freezing had yet to be formulated; instant freezing was not essential, although admittedly the enormous hauls that have since been made possible by stern trawl and purse seine, especially of pelagic species, have created problems on freezer ships in recent years.

Northland

The example set by the Anglo-Norwegian mother ship enterprise of Hellyer's was once more emulated before the UK gave up the idea of mother ships for good. In 1930 Northland Ltd was established as another multinational operation; there were three directors of Associated Fisheries Ltd, the group built by William Bennett of Grimsby and Billingsgate; Irvin's of North Shields and Aberdeen were represented, there was some Norwegian investment, and it was also claimed that Harden Taylor's company, Atlantic Coast Fisheries in the USA, had a stake in it. Whatever the exact composition of the Board, the complexity of management led later to a falling out among directors, and reconstruction of the company in a new guise.

The company was reported to have bought a 10 000-ton ship, *Highland Enterprise* (although one contemporary report quotes her as being 5165 registered tons, and another names her as *Highland Pride*) and after fitting her out with freezers and cold stores, renamed her *Northland*. She was reported to have cost £20 000 to buy, and a further £60 000 to convert. This time there were no dories; she was to be accompanied by a fleet of UK steam trawlers and liners, some of them acting as carriers, and by a 4340-ton collier called *Kamir*, which was to carry fuel on the outward passage and fish on her return.

Her brine immersion freezer was said by Zarotschenzeff to be a Piqué design, but it may simply have been that the Ottesen plant referred to in other reports was yet another brine freezing installation that Piqué had advised upon, as he had done for Hellyer's ships. The *Northland*, with a Norwegian, Captain Thorsen, in command, left North Shields accompanied by six steam trawlers and six steam liners towards the end of May 1930; the mother ship went into Methil to take on coal, to Norway to pick up frozen bait, to Faroe to gather more vessels around her, and thence to Greenland, but the trawlers and liners made their own way there direct. The first trawler carrier run was made by the *Ben Meidie* of 1928 fame, arriving in Aberdeen with 50 tons of halibut on 21 June; the fish had been transferred frozen from mother ship to trawler,

34

but allowed to thaw on the passage home. *Ben Meidie* was back again, this time to Grimsby, on 28 July with a second consignment; Markham Cook, as an Associated Fisheries company, handled the catch in Grimsby. But when a third consignment of 34 tons came into Grimsby on the steam trawler *D W Fitzgerald* in August, Mr T T Irvin arrived with her, and all was not well. As a director of Northland he had sailed with the mother ship as manager, but had returned early and promptly severed his connection with the company. From then on, since most of the trawlers and liners were owned by Irvin and other Shields and Aberdeen owners, each vessel was released from contract after completion of the first run home to market. The last of the North Shields vessels to arrive at her home port was the *Ethel Irvin*, which had been designated flagship when she set off; she came back on 1 November. The collier *Kamir* brought 600 tons of salt cod back to Aberdeen in October. The first season of the *Northland* had been a failure; there were said to be manning problems and mechanical difficulties, but bad management was an important factor. There was an unconfirmed report that *Northland* was working off West Africa late in 1930, and landing fish at Las Palmas, but by 1931 Bennett and the Norwegians had assumed control of a reconstructed company, and renamed the ship *Thorland* after some conversion work in readiness for the 1932 season.

Thorland

The *Thorland*, formerly *Northland*, made her first five-month Arctic voyage for halibut accompanied by eight ships of the London Whaling Company! No explanation of their role in the fishery was given, but in the spring of 1933 *Thorland* was still dispensing frozen halibut from Grimsby, having lain there since her return in October 1932, so presumably a reasonable catch was made. It seems almost certain that the *Thorland* had reverted to the use of dories, and that the whalers served as local carriers between line fishermen and mother ship. A second six-month voyage was completed in 1934, but when *Thorland* returned to Royal Dock, Grimsby, in October she discharged her frozen catch to shore cold storage there, and left immediately for Yarmouth to freeze herring. Lumley and Piqué were asked by Bennett's of Billingsgate to design a wire mesh drum immersion freezer especially for use with herring and, with the new equipment on board, the *Thorland* worked day and night at Yarmouth Quay from 31 October 1934. Herring were brought to the ship by lorry or direct from steam drifters as they arrived and the frozen herring, produced at the rate of 32 tons a day, were transported partly to cold stores in Grimsby and partly direct to canners and kipperers.

Thorland had only one Yarmouth season apparently—the idea was quickly dropped because of rancidity and other problems with the finished product. The *Thorland* continued her summer voyages to Greenland, however, and was last reported arriving at Grimsby with frozen halibut in October 1937, having spent six months supplied by Norwegian-manned dories. She was thus the last of the British mother ships to remain in service, 11 years after *Helder* set off in the spring of 1926. This long and no doubt unprofitable series of experiments had included fishing by dory, steam trawler and steam liner, the use of Norwegian, Faroese and Newfoundland fishermen, and the employment of refrigerated carriers, colliers and whalers, but the right combination to serve a high seas factory expedition had eluded the owners; there probably was no profitable solution.

Volkswohl

There were no commercial freezing installations on British trawlers between the wars. Lever Brothers came near to one in 1930, when they pored over outline plans for a 200-foot freezer trawler with Piqué in attendance as the DSIR brine freezing consultant but, since they had close links with Nordsee in Bremerhaven, they decided to await reports on the government-backed German experimental ship *Volkswohl* then just reaching the sea trial stage.

In 1926 the German government made available £30 000 as a state grant to build an experimental freezer trawler. The 52-metre ship was constructed by Deutsche Werke at Kiel, and ran her builder's trials in December 1929. She did about 10 knots, had an endurance of about 45 days on fuel oil, and carried a crew of 22 including three people to run the freezing plant. Auxiliary diesel engines provided power for an electric winch and for the refrigeration machinery, which consisted of Borsig carbon dioxide compressors to cool brine to −18°C. Ottesen brine immersion tanks with a total freezing capacity of 15 tons in 24 hours were in the form of six double containers holding twelve

The first purpose-built freezer trawler, *Volkswohl*, built in Germany, 1929

150-kg batches of fish, with deck loading hatches above. Brine was pumped in at the top of the tanks and drained off at the bottom when freezing was complete; fish were removed, washed, glazed and stored at about −20°C, in two cold stores, one forward and one aft, with a total capacity of 120 tons. She was unique in that she was the first trawler built new for freezing at sea, as opposed to conversion. Fishing trials began in the North Sea in January 1930, and were followed by an Iceland fishing trip; she returned in early March having caught only 80 tons of fish, of which 35 tons, mainly cod, haddock, redfish and saithe, were frozen. After a second voyage and another 35 tons of frozen fish, she went off, like almost all other freezer ships, to try halibut freezing off Greenland throughout the summer.

At the end of 1930 she was declared a technical success, but so far had proved uneconomic, mainly because retailers and consumers in Germany lacked interest in frozen fish. It was proposed to make a number of alterations, for example to provide chilled buffer storage below deck, to do away with the open deck hatches which had proved dangerous in bad weather, and to change from the batch tank system to continuous freezing, but before these plans could be put into effect the money ran out; government funding of the experiment was withdrawn and the ship was laid up by the end of 1931, victim of the depression and the absence of commercial support.

Norwegian ventures

Apart from joint participation in the *Helder* and *Northland* expeditions from the UK Humber ports, Norway also made at least two other attempts to freeze fish at sea in the 1930s. The steamship *Alekto* from Tonsberg was equipped with a Sabroe freezing plant in time for the 1930 herring season; capable of handling 7 tons in 24 hours, and with a 620 cubic metre cold store, she presumably served only as a processing ship and not as a catcher, but nothing is known of her performance. The *Korsvik* from Oslo on the other hand followed the pattern of the *Arctic* mother ships; equipped with J & E Hall carbon dioxide compressors, a brine immersion freezer to handle 10 tons a day and storage for 500 tons of fish at −18°C, she followed the familiar route to Greenland in the spring of 1931 and returned in September to Grimsby with a cargo of frozen halibut. Whether she continued to operate in successive seasons is not known.

Jean Hamonet and *Marie Hélène*

CAFPA, owners of the *Sacip* and concessionaires of the process of that name, were sufficiently satisfied with the performance of the segmented drum freezer to plan a second installation. The oil fired steam trawler *Jean Hamonet* of La Rochelle, 52 metres in length and built in 1927 as a salting trawler, was fitted out at Bordeaux in 1931 with carbon dioxide compressors, two drums that together froze ¾-1 ton an hour, and a cold store. The drums, each 2.8 metres in diameter and 1.4 metres long, had eight segments, the lower four of which were submerged in brine at −20°C. Each segment held 300 kg of fish, and freezing times were claimed to be 2½ hours for large fish down to 40 minutes for small ones.

She made her first trip as a freezer in the summer of 1931 and, after fishing off Newfoundland, paid a call at Boston in August to demonstrate her plant before returning with about 50 tons of frozen fish. She may also have had salt fish on this trip, but on a succession of voyages to Iceland, Newfoundland and the West African coast she often made comparatively short excursions for freezing only; for example she returned from Mauretania on one occasion with 130 tons of frozen

The brine drum freezer fitted on the French trawler *Jean Hamonet* in 1931

SECTIONAL ELEVATION.

fish after only 10 days fishing. She was said to carry an 'unnecessarily large' crew of 30 to comply with certain French regulations, and was away for periods ranging from 18 to 38 days, a pattern totally unlike the conventional salting voyages. Two factors that contributed to her success were the use of shore cold storage to permit rapid turnround, and the markets found for her product; fish was dispatched frozen from the port cold stores and was allowed to thaw en route, the biggest customers being the French armed forces, institutions supported by public assistance, and other state-supplied groups. Obviously this kind of catering outlet, through state contracts to consumers who had little opportunity to choose, bypassed the conventional routes through distributors and retailers, and the resistance often encountered there. Nothing is known about quality, but it would be reasonable to assume that, with a steady demand and regular supplies, residence time in cold storage was probably short, and that eating quality was no worse than that of any other brine frozen fish of the period. Results were satisfactory enough to make the company equip a second ship almost right away; the *Marie Hélène* was fitted with an identical freezing installation and put into operation in 1932. After 4 years working with two trawlers, another two bigger and better ships were added to the fleet in 1936, as described later.

Lumley and the UK owners

After the humble trial on the *Ben Meidie* in 1928, Hardy had installed Lumley as Superintendent of Torry Research Station in Aberdeen in 1929 to lead a small team engaged on more systematic work on the problems of freezing and cold storage of fish. Piqué stayed at Cambridge, but remained in close touch with what Torry was doing, since both he and Lumley remained desperately keen to see the innovation of freezing in the British trawling industry. Piqué addressed the professionals at the British Association of Refrigeration in 1931, Lumley wrote for the technical and popular press, having made some studies of the likely economics of freezing part of a trawler's catch, and they were both in frequent communication with leading lights in the industry. They tried the first of many public showings of frozen fish in 1932, when a press demonstration was put on in London, using fish that had been brine frozen on shore soon after capture by the station's research vessel, to simulate the sea frozen product; the programme 'was designed to resemble as closely as possible . . . brine freezing and storage of fish aboard a trawler' but the reporters who attended at DSIR headquarters put very little copy into their editors' hands.

Two refrigeration companies, J & E Hall and L Sterne & Co, prepared layout drawings and estimates of cost for a Piqué brine freezer in a trawler, so that Torry staff could present potential clients with detailed information, but all to no avail. In March 1933 Lumley wrote to Hardy saying that he had touted the plans for freezing the first part of the catch around all the principal owners in Hull, Grimsby and Aberdeen, but none had shown any interest in investment. Since Lumley believed, correctly, that a mother ship was not the answer, and that the French supertrawlers to Newfoundland, though apparently successful, were not applicable to the UK fishery, he therefore argued that government and industry must jointly finance a demonstration of freezing the early part of the catch by chartering a trawler for about a year. Although his proposal was turned down flat in those financially stringent times, some 20 years later government and industry did exactly that.

One of the few trawler owners who set down their reasons for declining Lumley's propositions at that time was Andrew Lewis of Aberdeen. He was a trawler owner, a shipbuilder, and a former Lord Provost of the city in which he had many financial interests; indeed he had been instrumental in establishing Torry 4 years earlier. As one of those in the fish industry least likely to be short of investment capital, his letter of refusal is pertinent:

> 'Dear Mr Lumley,
> I had a discussion with my associates about your brine freezing proposals with the result that we came to the conclusion not to go in for it, at any rate at the present moment because of the financial risk . . . There are two ways in which it may not succeed. First, there is the unknown difficulty of operating the troughs (freezer drums) at sea, and other possible mechanical difficulties. Second, there is the uncertainty of there being a market for the brine frozen fish. If times were better we might take a chance, but as things are we dare not.
> It is not as if this particular scheme were already in operation and we were reasonably sure of what the snags would be, but it is really a case of being the first to venture, and you know the pioneer usually loses his money.
>
> > Yours truly,
> > Andrew Lewis'.

It has to be admitted that his fears were to some extent justified; the popular image of frozen fish was still a poor one, brine frozen fish were sometimes far from attractive after long term cold storage, and George Reay's advocacy of much lower storage temperatures to alleviate this problem had still to come.

Oscar Dahl's views on freezing at sea

Dahl, who had demonstrated his K-D system on the *Pen Men* in 1927, is today perhaps best remembered in the fishing fraternity for his pioneering of the Vigneron-Dahl, V-D, trawl gear in the mid 1920s, but in 1933 he was asked by the British Government to give his views on the relative merits of commercial freezing systems then available; obviously he was biased in favour of his own method, but his comments on his rivals are worthy of note.

He accepted that the SACIP system, by then installed on *Jean Hamonet* and *Marie Hélène*, was compact and easy to operate at sea, but considered its first cost to be too high for the average trawler. He saw it as a direct derivative of the Piqué, Ottesen and Kolbe immersion systems using mechanically cooled brine. He claimed that the *Jean Hamonet*, designed to freeze 20 tons a day, could in practice achieve only 9 tons, and that both capital and running costs were unacceptably high; the fact that her owners subsequently converted two more ships does not seem to support that view.

On the question of which direction future shipboard freezing installations should take, he was still arguing against the need for low temperature storage, and in favour of what he called light or short freezing. He believed that the catch should be disposed of quickly after landing, and that total time in cold storage need be no longer than 25-30 days, and therefore that the pattern of events should be freezing in a refrigerant at −12 to −14°C to take the fish down to about −5°C, and then storage pending distribution at −3°C. To achieve these conditions, it was not necessary to install compressors and mechanical briners that could so readily fail in the harsh operating conditions that prevailed at sea; the good old ice-and-salt freezing mixture, with nothing more complicated than a pump, was cheap to install and run, and was adaptable, simple and foolproof. He challenged the view that this was only superchilling, or worse, just chilling, and claimed that for his purpose of short term cold storage, it worked. Since the *Pen Men* days he had fitted out a trawler called the *Hourtin* which had been demonstrated to the hake trawlermen at Milford Haven, but it had subsequently sunk off the south coast of Ireland. Undeterred, Dahl was fitting more of his fleet at La Rochelle.

Although Dahl's was only one voice in the seemingly endless discussions about how to freeze at sea, it helped to sow seeds of doubt in the minds of those in high places in the British fish industry—should one encourage and, more importantly, put money into mother ships, freezer trawlers, superchilling trawlers or, as was soon to be considered, fillet freezing trawlers? While ever increasing amounts of poor quality iced white fish from distant waters continued to appear on the markets at British east coast fishing ports, government and industry hesitated—and did nothing more about freezing at sea for the rest of the 1930s.

Fismes and the B-F system

After a long series of small scale experiments, a variant of the Dahl method was installed on a ship from La Rochelle to work off West Africa in the summer of 1934. The Bellefon-Folliot system, B-F, as used on the *Fismes* consisted of the familiar brine trickle at about $-3°C$, but this time the fish were in watertight steel boxes that held about 50 kg, thus seemingly making it unlikely that anything more than chilling would be achieved, if that. Nevertheless claims were made of a 40-day shelf life for the 100 tons of fish that were brought back and distributed as chilled, not frozen, fish. The patented system was offered for sale in the UK by René Moreux & Cie, Paris, but there were no takers.

Dry ice

Blocks of dry ice, that is solid carbon dioxide, marketed under the brand name 'Drikold' by Imperial Chemical Industries, were tried as an expendable refrigerant on a Grimsby trawler in 1934. The method of use was probably much the same as was tried in railway wagons and other land transport; blocks would have been stowed above the cargo in the fishroom once stowage was complete, possibly in a perforated container of some kind just below a deck hatch, so that cold air could flow down around the catch. Much the same practice was revived for a short time in the late 1940s and early 1950s, but again proved ineffective. Low temperatures in the immediate vicinity of the dry ice usually resulted in some partial freezing that made the fish unattractive in appearance, whilst the greater part of the catch, surrounded by ice in the conventional manner, remained completely unaffected by the presence of the ancillary cold source. Whatever the exact details of the 1934 Grimsby experiment in 'freezing' with dry ice, it was immediately declared a failure and abandoned.

Vivagel and *Pescagel*

These two French trawlers continued the SACIP tradition when they were equipped in 1936. Another two installations were also planned—*Vivagel* was originally described as the first of four—but they never materialized.

Vivagel, 65 metres long and described, when built in 1927, as the world's largest trawler, was fitted out by May 1936 with bigger and better SACIP segmented drums that could freeze 1½-2 tonnes an hour; brine at $-20°C$ was cooled by carbon dioxide compressors and was pumped through the hollow axles of the long drums that lay fore and aft below deck, one on each side of the ship. The cold stores had a capacity of 300 tonnes at $-18°C$, and she returned in September almost full from her maiden trip to Newfoundland. The fish went into the SACIP company's cold stores at Boulogne and Marseille for distribution. *Pescagel* was equipped in like manner by late 1936 or early 1937, and they continued to operate satisfactorily, apparently, until the outbreak of the Second World War. Whatever the economics were of

The French trawler *Vivagel*, fitted with
brine freezers in 1936

this venture, it was the only one in the years between the Wars that
managed to freeze fish at sea on ships that caught it, as opposed to
mother ships, and kept going for the best part of a decade until
interrupted in 1939. Perhaps it survived only because of a series of
institutional catering contracts from a supportive government, but
those French supertrawlers as Lumley called them, dashing to and fro
across the Atlantic to work the still prolific Newfoundland Grand
Banks, were all there was to show for over 20 years of shipboard
experimentation when war broke out again.

Freezing fillets at sea

Hand filleting of fish at sea had always seemed such an uninviting
prospect in the hostile environment of a rolling trawler that freezing of
fillets at sea had never been given much serious consideration. But
prototype filleting machines were beginning to make an appearance,
and in 1937 when a model from Harden Taylor's company in the USA

was being publicised, Torry gave some thought to the possibility. The obvious advantages were the improvements in speed and output from freezing when the thickness of the commodity was reduced from that of whole fish to fillets, and increase in shipboard storage capacity by reducing bulk, thus improving the ratio of fishing time to unproductive time of the ship. Against these had to be weighed the disadvantages of making a product that was more limited in its use, and about which little was known in regard to its behaviour during freezing and cold storage.

With *Vivagel* in being as a trawler freezing whole fish for the French market, it was suggested in the UK that a fillet freezing trawler would provide useful comparisons; Torry Research Station was party to a proposal that a ship should be designed and built by a reputable trawler builder and equipped by a prominent refrigeration company; Smith's Dock and L Sterne & Co respectively had been canvassed and were willing to fulfil these roles but—and it was a big but—Government was expected to put up the money, and there the dream ended.

Lower storage temperature

George Reay wrote in late 1936, shortly before he took over from Lumley as superintendent at Torry, about how the continuing poor image of frozen fish could be improved:

> 'For many years freezers of fish in this country have been . . . keeping their fish for some months at temperatures ranging from $14°$ to $20°F$ (-7 to $-10°C$). While such fish remain perfectly fresh in the sense that bacterial activity has been eliminated, they are so distastefully unlike fresh caught fish that they are largely responsible for the prejudice which exists against frozen fish.'

Going on to explain that the fish suffered from freezer burn, rancidity, poor appearance and a cooked texture that was dry and woolly to the palate, he gave the remedy. 'The secret is to keep the temperature of storage as low as -10 to $-20°F$ (-23 to $-29°C$). . . . The fish are every bit as good to eat as freshly caught fish. Considerable engineering research will be necessary . . . to discover how best mechanically to handle large quantities of fish and to freeze them with sufficient rapidity . . . aboard as soon as they are caught.' Having linked together, perhaps for the first time, the three requirements—newly caught fish, quick freezing to a low temperature, and maintenance at that temperature—he went on to consider the seaborne options for catches on remote grounds. The factory ship or mother ship supplied by catchers faced the insuperable obstacle of transshipment in rough weather; interruptions in supplies prevented round-the-clock operation and made the venture uneconomic. 'One would like to see a ship specially designed' wrote Reay, 'for both catching and freezing. Nothing has been done on those lines in this country as yet, but a French company has several such ships.' But Reay had to wait till after the War to resume his campaign and see his wish fulfilled.

Hamburg and Weser

Almost as a tailpiece to this section are two German wartime experiments that came about largely as a means of augmenting scarce food supplies rather than as a means of improving quality and extending storage life.

The *Hamburg* was a 5500-ton ship owned by Andersen of Hamburg that was fitted at government expense with a continuous belt air blast freezer in 1940. The refrigeration is believed to have been a flooded ammonia system with a surge drum; air was cooled to $-45°C$ and blown over the fish until surface temperature was down to $-22°C$ before storage. Freezing capacity was 50 tons of fillets a day in the form of blocks in trays, and there was storage for 800 tons. Fish meal and oil plant was also installed. It is unlikely that this ship was ever intended for work at sea; she was meant for operation in Norway, probably at anchor in a fjord or even alongside the quay in a port. Whatever the intention, a fortnight after her arrival at Svolvaer in March 1941 she was sunk in a British air raid.

The *Weser* began life as a conventional steam trawler of about 61 metres in length that was sunk in 1942, raised, lengthened 2-3 metres and equipped with filleting, freezing and byproduct plant; the conversion, which was heavily subsidized, was meant to provide a ship that could process fillets in the comparative safety of the Baltic for onward carriage to Germany. Fish were delivered by chute from the fishing deck to a Baader filleting machine below; the fillets were made up into 1 kg blocks, wrapped in parchment and put on trays. The trays were laid on coiled pipe shelves which formed the basis for a hybrid air blast—contact freezer; the piping was cooled by direct expansion of ammonia, and cold air was blown over the fillets by fans. The freezer was apparently based on a fruit and vegetable freezer patented in the UK by Heckermann, and similar in principle to another design known as the Murphy freezer. Final adaptation of the plant for the *Weser* was by a Dr Schlienz. The ammonia evaporation temperature was $-32°C$, the freezer could handle 1.6 tons of fillets from 4 tons of fish an hour, and the blocks, packed in cartons, were held in a brine cooled store at $-20°C$.

The *Weser* worked in the Baltic for about 2 years from 1943 to 1945, and she was seen by George Reay, in Wesermunde, just after the War ended, in the course of a tour of inspection of the German fish industry; at that time, early 1946, the freezing plant was still aboard, though partly stripped, and those who had managed her operation said that she had worked reasonably well. She could fish for her own supplies in winds up to force 7, and produced a satisfactory frozen product, but she had never shown a profit, and was too small; the space that remained for storage after processing plant was installed was inadequate. In spite of this, Reay thought the *Weser* was a useful pointer, and advised British industry to study the experiment closely.

In the immediate postwar years the fishing fleets of Europe were busy catching fish; after several years' respite from overfishing, the stocks in the north east Atlantic had had time to replenish, and hordes of weary veteran trawlers released from war service quickly earned enough money to help their owners build new replacement tonnage. But soon there were fine new ships making long trips and having to come home partly full, in order to land the catch before it became completely spoiled. As food supplies increased and the consumer again had a choice, fish of indifferent quality was spurned, and catchers once again had to concern themselves with acceptability as well as quantity if they wanted to stay in business.

Ultimate success in the art of freezing fish at sea was to depend a great deal on the catchers' awareness of the need and on their eventual financial commitment. But in the postwar years there were other important contributory factors that made the difference between success and failure; mistakes were still to be made of course, and a number of costly experiments came to nought, but the odds in favour of a satisfactory solution were greatly enhanced by developments in three important areas. First, there was now a small group of people who, after years of systematic study, were able to specify precisely the conditions of freezing and cold storage that would yield a product virtually indistinguishable from fresh fish; Reay and others had already begun to spell out those conditions just before 1939—quick freezing, much lower storage temperatures and adequate protection against dehydration and rancidity. Secondly, there had been significant advances in refrigeration engineering, especially in the range of refrigerants available for shipboard use; halocarbons, first used as refrigerants about 1933 in the USA, were rapidly exploited after 1945 and were used in machines that were safer, more compact and more efficient. Thirdly, there was the appearance on the scene of a new kind of engineer who could take the components of landbased processing equipment and, having acquired an intimate knowledge of the constraints of the fishing vessel and the special requirements of fish as a frozen commodity, adapt and complement them to make a robust and efficient shipboard system professional engineers like Lochridge, Eddie and others of that ilk emerged to convert what the refrigeration engineers had to offer into what the trawler owners and fish technologists wanted.

47

Sir Charles Dennistoun Burney was an inventor of renown; a former naval commander, he had developed the paravane for minesweeping in the First World War, and in the 1930s had been active in the promotion of airships. In 1945, while working with ICI at their factory at Ardeer in Ayrshire, he first came up with the idea of a fleet of 20 fishing vessels that would produce frozen consumer packs, for sale not at the traditional fishing ports but at ports like Glasgow and Manchester, where the buyers would be large chain store groups and other retail outlets.

The first task was to make a freezer, and then to acquire a ship on which to develop a suitable prototype installation; to this end a company called the Broadway Trust was set up, and by early 1946 a pilot freezing plant was under construction on the premises of L Sterne & Co in Glasgow. Burney revealed his plans to Reay and his colleagues from Torry in February 1946, and sought their assistance in testing the prototype. Torry staff were present during the first test run of the Burney freezer in May 1946; fish from Torry's research vessel were taken to Glasgow for the event. Reay likened the freezer, part air blast and part direct contact, to the one he had inspected on the *Weser* and to the Murphy freezer, details of which seem to have vanished. The trial unit consisted of an insulated chamber containing shelves made of parallel lengths of tubing, finned underneath, on which trays of small whole fish or wrapped 14 pound blocks of fillets were laid; cooling was by conduction to the finned tubes and by fan circulation of cold air. Burney believed that in this form the freezer could be easily operated by unskilled crew on a rolling ship, and would occupy the minimum of space. Torry staff attended three tests in all and, in addition to giving instruction on the measurement of freezing times and other aspects of performance, suggested to Burney several improvements in design.

Any fishing vessel that was to incorporate a factory deck as Burney envisaged would of necessity be a highsided ship, from which conventional side trawling would be impracticable. Burney and his colleagues came to the conclusion very early on that stern trawling, an art that had been abandoned with the demise of the beam trawl in the late nineteenth century, would better suit their needs, particularly since Burney had ideas of adapting his paravane technique to the trawl. While the freezer was being built in 1946 the company acquired the steam yacht *Oriana* on which to try out their revolutionary trawling ideas; her stern was modified, and a series of trials with the parotter, Burney's paravane replacement of the conventional trawl otter board, was conducted in the Clyde estuary. Having satisfied himself about stern trawling, and having a freezer that seemed to work, Burney needed to put them together on a ship and go fishing, but he could not

48

afford to wait for delivery of new tonnage; the queues were long at shipyards in the immediate postwar years.

At this point the relationship between Burney and Torry became somewhat strained; he was about to publish a booklet advertising the Burney freezer, and in order to assist in fund-raising for the shipboard venture he wanted to declare that the freezer was 'approved' by DSIR, as a consequence of Torry's attendance at the shore trials. This cachet was denied him, and he apparently broke off diplomatic relations. Early in 1947 his company, now called Fresh Frozen Foods Ltd, bought a secondhand 1500-ton Algerine class minesweeper, built in Canada and named HMS *Felicity*. By this time Burney had working with him a Clydeside engineer named William Lochridge, who had previously co-operated with Burney at Ardeer on the design of a recoilless gun, and Lochridge took over most of the design and installation work for the conversion of this ship and the eventual construction of her successor. *Felicity* became *Fairfree*—she was meant to be *Fairfreeze*, 'Fair' in recognition of Fairfield's shipyard where she was to be remodelled, and 'freeze' to indicate the object of the exercise, but those in the Board of Trade who had jurisdiction over the naming of ships disallowed the latter part on the grounds that it sounded plural and might be misconstrued.

In the summer of 1947 the *Fairfree* was equipped with an improved version of the Burney freezer which became known as the Fairfreezer; the trays of fish now sat on knife-edged fins on the upper faces of the piping shelves, so that cooling was almost entirely by moving air. The secondary refrigerant in the pipes was a brine at about −28°C, cooled by ammonia compressors. The cold store was run at −20°C and could hold about 200 tons of cartoned fillets. The intention was to produce two sizes of block, a small 2 pound consumer block, and a 14 pound one

Stern trawler *Fairfree*, equipped with freezing plant in 1947

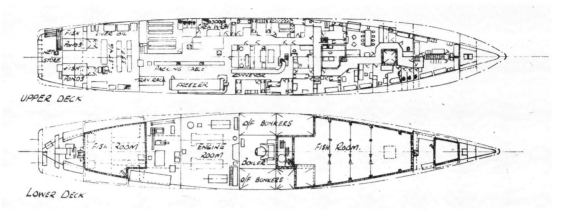

UPPER DECK

LOWER DECK

Fairfree: plan view of upper and lower
decks

wrapped in cellulose film and packed in fours in an outer carton.
Freezing capacity was about 1 ton an hour.

Fairfree was shown off proudly to the press at Ardrossan, where the
final conversion work had been done, on 24 October 1947, and she was
put through her paces in the Clyde estuary before the end of the year.
Fishing trials were conducted in the North Sea early in 1948, and then
it was decided to convert her to diesel propulsion before tackling more
distant waters. Her first long voyage was to Newfoundland later that
year, and there were some teething troubles; the filleting machine for
example proved unsatisfactory at sea, and hand filleting had to be
resorted to. From then on whole fish were frozen, except when filleting
was necessary to make them small enough to go into the freezer. She
continued to work spasmodically, and lessons were learnt, but she
remained very much an experimental ship.

In the autumn of 1948 Burney approached Christian Salvesen, a
whaling and merchant shipping company in Leith, partly because of
their experience with stern ramp operations on whale factory ships,
and their interest in the freezing of whalemeat; the outcome was the
purchase by Salvesen of Fresh Frozen Foods Ltd in May 1949, and the
transfer of *Fairfree* from a Glasgow fishing registration to a Leith one:
GW19 became LH271. Lochridge was part of the deal, and he remained
with Salvesen until 1955. *Fairfree* made a voyage for Salvesen in
autumn 1949 and continued to operate intermittently until September
1950 when she was laid up at Leith. By this time she had fished
Newfoundland, Norway, Barents Sea and Faroe in her 3 years of
operation, and continual improvements had been made, but she was
cramped, and was a poor performer as a trawler; Salvesen's decided
to continue the experiment on a new purpose-built ship. The
Fairfreezer was patented and the design, used again on *Fairfree II*, as
the *Fairtry* was first known, was also sold to the owners of *Mabrouk* in
1951. The stories of both these ships are recounted later. *Fairfree*

herself lay in Leith, unloved, for many years until she was finally towed out for scrap about 1962.

North American ventures

From 1944 onwards a number of ships were equipped with freezing plant of some kind, mainly for work in the Pacific or the Gulf of Mexico, but most of these are given no more than a passing mention here because the purposes they served bore little relationship to European operations and, for many of the ships, detailed information is not readily available.

A ship named *Soupfin* had a facility for the sharp freezing of fillets when working from the Pacific coast of the USA in 1944, and the Gulf shrimping fleet had an experimental mother ship called *Betty Jean* in 1945 that operated in collaboration with Louisiana State University, and was described as the 'first floating shrimp packer and freezer'. The *Chirikof* was reported to be working from Seattle to the north Pacific fishing grounds in 1945, and was said to be filleting, packing and freezing fish ready for the consumer, in addition to freezing crab. The *Alaska Queen* and *Aleutian Queen* operated in the same area, but nothing more is known about them. In 1946 the *Helen Crawford* was claimed to be the only floating, freezing and packing plant in operation at that time in Alaskan waters, so the implication is that many of these

The USA trawler *Deep Sea*, fitted with an air blast freezer in 1947

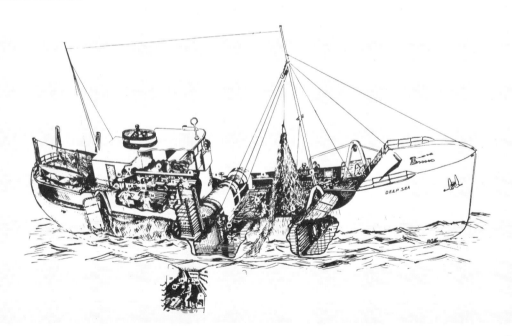

shipboard installations never worked for much longer than a single season. In 1947 the motor trawler *Sea Horse* was reported to be freezing packs of fillets at sea, possibly off Florida.

The *Deep Sea* from Seattle is somewhat better documented. She was 140 feet long, 350 gross tons, and was designed to catch and freeze mainly king crab, but also to catch other fish for filleting and freezing. The catch was conveyed mechanically from fishing deck to storage hold, and was blast frozen in transit; fillets were packed in 36 pound lots in pans for freezing, and storage temperature was −18°C. She had storage for 150 tons. Her first voyage was to Alaska in the summer of 1947, and in early 1948 one reporter said that it was not known whether she had been a success, which perhaps suggests that she had not.

The *Pacific Explorer* was a big mother ship after the pattern of *Helder* and *Northland* in the interwar years, but less successful. She had been built as a bulk carrier during the First World War, and was far from ideal for conversion, but was all that was available when funds were provided by the Defense Plants Corporation, a subsidiary of the Reconstruction Finance Corporation, in the last months of the Second War. She was operated by the Pacific Exploration Company, and was meant to be all things to all fishermen, with dire results. She was 410 feet long, with seven freezers, three air blast and four 'shelf-type', which were to handle between them 160 tons of fish a day. Seven cold stores ranging from −18 to −23°C had a capacity of 2350 tons. Her primary purpose was to work in the Bering Sea with about 240 people on board, while a secondary objective was the tuna fishery in the southern Pacific, with a crew of about 80. It was reported afterwards that her dual role was the main cause of her failure; the tuna fishery wanted brine immersion freezers, which she did not have, and the northern fishery could not meet the demands of her huge freezer capacity. Furthermore, she was far from trouble free; her two-stage flooded ammonia system was declared to be too complicated for shipboard operation, and was frequently out of action. She went into service in 1947, but it seems unlikely that she worked in succeeding years.

Towards the end of 1948 the US Fish & Wildlife Service began modest investigations into freezing at sea off the Atlantic seaboard, using their biological research vessel, *Albatross III*. She was fitted with a small cold store running at −29°C in which small lots of whole ungutted fish were sharp frozen in still air; the store temperature rose to −23°C when it was filled. The frozen fish were boxed and stored at −18°C for subsequent work on shore; trips were typically only 5-10 days duration. The main purpose was to compare quality of fillets taken from frozen fish with those from iced control samples. The results were sufficiently encouraging to initiate further work using brine immersion freezing, with commercial freezing at sea as the ultimate goal. Brine freezing was considered to be the only option, because still air was too slow, air

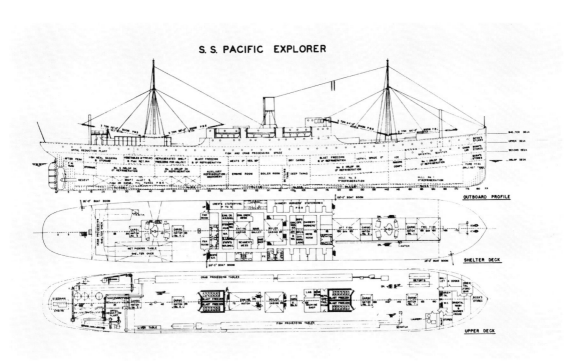

S. S. PACIFIC EXPLORER

Mother ship *Pacific Explorer*, fitted with freezers in the USA, 1947

blast freezing took up too much space on the ship, and contact freezers suitable for seagoing work were not available. 'It is possible that a vertical loading mould freezer could be designed which would require a minimum of handling, but the equipment would probably be quite costly and complicated.' On the other side of the Atlantic Eddie and his colleagues at Torry were about to tackle just such a project, a simple and inexpensive vertical loading contact freezer that would work at sea.

UK options: guidance from Reay

George Reay, when addressing Humber trawler owners at Grimsby in 1951, outlined the options for the distant water fishery as seen by Torry. First, freezing at sea in some form or other was the only answer to the current problem of large trawlers spending more than half their time steaming instead of fishing, only to return half to two thirds full with fish of low quality; to secure an increase in both productivity and quality, the only process that arrested spoilage without markedly changing the product was quick freezing and low temperature cold storage. If they accepted the premise, then the choice was between a factory ship with attendant catchers and a ship that froze her own catch; Reay was adamant that the freezer trawler was the right solution. The complex, expensive factory ship that either lay in shelter far away from where the catchers were working, or lay idle on the high seas because bad weather made transshipment impossible, could never earn her keep. But there was room, even on existing trawlers, for a simple and compact freezer to preserve at least the first part of the catch so that fishing could continue till the ship was full. To keep the freezing plant small, some chilled buffer storage would be needed to cope with fluctuations in catching, and the success of such an installation depended very much on Torry managing to make a robust vertical plate freezer; work to that end was already going on, and shipboard trials on Torry's vessel were already under way. He discounted the idea of filleting before freezing; suitable machines were still not available, and hand filleting, difficult on a rolling ship, required too many extra hands. He also argued against freezing only high value species in small quantities; this did nothing to solve the problem of low productivity and poor quality in the distant water fishery for staple fish like cod and haddock.

Keelby

Eddie had begun work at Torry in 1948 on the design and development of the piece of equipment that was the key to Reay's freezer trawler concept, a vertical loading freezer for whole fish. Work had begun in the laboratory with a single-compartment freezer, first with the traditional ice can, both with and without water among the fish, and then with a variety of mould designs with hollow walls through which refrigerant, or hot gas for defrosting, could be passed. Once it had been established that good heat transfer properties were not dependent on the fish being in water, the container no longer needed to be watertight, and the walls could be separate from bottom and ends; thus evolved the idea of parallel refrigerated plates that could be moved if need be to release the frozen block; fish could be loaded between the plates from the top, and

Research vessel *Keelby*, which carried the prototype vertical plate freezer, 1951-53

the block released through the bottom or an end. The design principles for the single compartment were incorporated in a four-compartment version, and the four-station vertical plate freezer, as it became known, was ready to be tested at sea.

Torry's research vessel at that time was a 90-foot wooden motor fishing vessel called the *Keelby* bought secondhand, but hardly used, in November 1949. A tiny cold store was built in her fishroom in the summer of 1951, and the four-station freezer put inside it. There then began a long series of tests which had two aims, to assess and improve the freezer's mechanical performance at sea, and to determine what treatment of the fish was necessary to yield a high quality product, as measured by laboratory inspection of the frozen blocks. The *Keelby* made day trips, hundreds of them, out to sea just far enough to trawl

The four-station vertical plate freezer used on the *Keelby*

for cod, and back each evening after a couple of loadings of the freezer to land eight experimental blocks of frozen whole cod—gutted, ungutted, bled, unbled, loaded before rigor, or kept for a specified time in ice before freezing, until the right criteria had been established.

After 2 years of tests from 1951 to 1953, the experimental prototype was as right as it ever would be, and the rules for producing frozen fish of high quality had been formulated. But the freezer was still an experimental prototype; it required the touch of the industrial designer and the production engineer. In the *Keelby* version the refrigerant, one of the halocarbons, Refrigerant 12 (R 12), was distributed to the freezer plates through flexible hoses; plates were moved by pneumatic jacks, and blocks were discharged from the end of each compartment; all these and other features were to be improved upon in collaboration with the refrigeration industry.

The role of the *Keelby* as a floating test bed was over; an improved freezer would need to be demonstrated on a commercial trawler. No longer a miniature freezer trawler, she continued to bring supplies of fish for research until 1956, when she was sold. By 1960 she was working commercially in West Africa, and was subsequently lost off the coast of Cameroon.

Mabrouk

The Compagnie Marocaine de Pêcheries (CMP), had two trawlers working from Casablanca to the coast of Mauretania; by the time the iced catch reached port the quality was too poor for freezing on shore,

The French trawler *Mabrouk*, fitted with a Fairfreezer in 1951

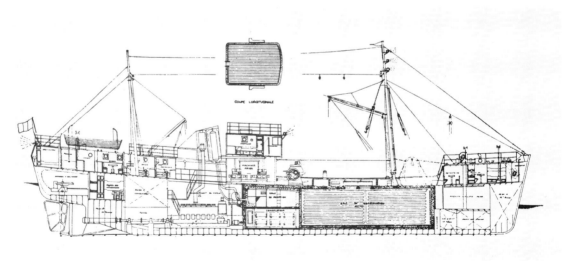

and the company decided to equip one of their ships, the *Mabrouk*, with freezing plant. They chose a Salvesen Fairfreezer, and installed it on the *Mabrouk* at Le Havre in 1951. They opted for freezing whole fish, arguing that there was insufficient space on their 50 metre trawler for filleting at sea, which in any event was difficult to do, and that there were wider outlets for whole fish. Their choice of a Fairfreezer was a process of elimination; brine immersion freezing made the fish too salty, and the stored product had a poor appearance, plate freezers available commercially were of the horizontal type as marketed by Birdseye and Jackstone and were suitable only for flat rectangular packs of uniform thickness, and conventional air blast freezers took up far too much space.

A freezing compartment was built in the after end of the fishroom, with a chilled buffer store above it that held 25-30 tons, enough for 2 days' freezing. An ammonia compressor supplied calcium chloride brine at $-35°C$ to the freezer pipe coils. Fish were frozen in blocks on trays, glazed, wrapped in cellulose film, cartoned and conveyed to a cold store that held 150 tons. The layout seemed to follow the Torry concept fairly closely, and Eddie commented '. . . the ingenious conversion shows what can be done, given a free hand. It demonstrates once again that freezing at sea has not to be developed once only, but anew for every fishery.' Off she went to work; by April 1952 her owners were saying 'results . . . were beyond our expectations', and in December CMP said 'we have been very satisfied . . . but people are still rather reluctant in buying frozen fish'—and then silence. Presumably yet another venture had failed, probably because of unreceptive markets.

Delaware

In the USA the initial experiments on *Albatross III* were being followed up by adaptation of a trawler to conduct further trials on a commercial scale off the New England coast. Government research workers at the East Boston laboratory were in favour of brine immersion as being best suited to New England trawlers that wanted to convert without extensive modification; new construction was not an option. The *Delaware* was 148 feet long over all and carried a crew of 20. She was fitted with a brine tank and a segmented drum, a direct descendant of the SACIP and Piqué freezers of prewar years, and comparatively high temperatures were accepted; fish came out of the freezer at $-15°C$, was stored in the 50 ton ship's cold store at about $-7°C$, and subsequently on shore at $-18°C$. Refrigeration was from an absorption system, on the grounds that it was cheaper and required less space. She began a series of experimental trips in the autumn of 1951, staying out for 8-10 days after the manner of her commercial counterparts. Her first landing of about 3 tons of cod and haddock was considered to compare favourably with iced fish. Experimental work on the ship and in the laboratory continued throughout the 1952 season, but there still

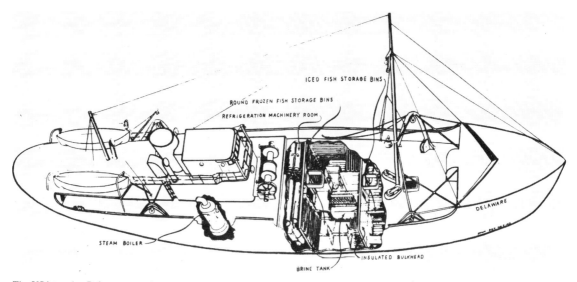

The USA trawler *Delaware* tested a brine drum freezer in 1951

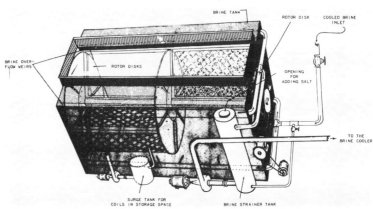

The drum freezer fitted on the *Delaware*

remained problems with plant operation at sea, and there were difficulties with thawing. Although storage temperature was high by UK standards, the associated problems of poor appearance of brine frozen whole fish at that temperature were not regarded as serious, because all of the fish were to be filleted and refrozen before distribution. This system of brine immersion, high temperature storage, and a second freeze after filleting, was one that by the 1950s would have met with strong disapproval in Torry as a solution for the UK industry.

Fairtry

Reay and Eddie remained firmly convinced that the right course for UK trawlers was the freezing at sea of whole fish; subsequent processing could be done on shore. But Salvesen's, after much heart-searching in the boardroom, decided to continue the good work done on the *Fairfree*, and in September 1950 approved the building of a new factory stern trawler, *Fairfree II*, that would fillet and freeze the catch at sea,

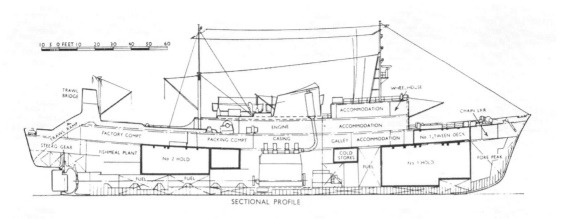

Labels within diagram:
TRAWL BRIDGE, WHEEL HOUSE, CHAIN LKR, ACCOMMODATION, TRAWL RAMP, STEERG GEAR, FACTORY COMPT, ENGINE, ACCOMMODATION, PACKING COMPT, CASING, GALLEY, ACCOMMODATION, No 1 TWEEN DECK, FISHMEAL PLANT, No 2 HOLD, COLD STORES, FUEL, No 1 HOLD, FORE PEAK, FUEL, FUEL, FUEL

SECTIONAL PROFILE

Fairtry, the first factory stern trawler, built in 1953

and make fish meal and oil from the waste; when she was eventually launched from the yard of John Lewis & Son in Aberdeen on 30 June 1953 she was named *Fairtry*.

With an overall length of 280 feet, and 2605 gross registered tons, she was far bigger than any other UK trawler of her time. Sterne's built her three Fairfreezers, and supplied compressors that ran on R 12; she also had horizontal plate freezers of standard design, and could freeze about 30 tons of fillets a day. Her cold stores could accommodate 600 tons of fillets at −20°C. She also carried a Sabroe icemaker to supply ice for buffer storage, and fish meal and oil plant to utilize the offal. She could carry 75-80 people, and was expected to make trips of about 3 months' duration. She could steam at 12-13 knots, and was ready for her first sea trials in April 1954. *Fairtry* was a first in a number of ways; she was the first commercial UK trawler to be built with a stern ramp for trawling—*Fairfree* had been a makeshift conversion. She was the first UK trawler built to freeze her own catch—her few experimental predecessors had been conversions from existing tonnage— and she was the first UK ship to freeze fillets commercially at sea, since her forerunner, *Fairfree*, had abandoned the attempt almost as soon as she had begun.

She was not an immediate unqualified success, as Lochridge and others admitted, but she was by no means a failure. A first class skipper proved to be essential if fish were to be found and captured fast enough to keep her factory working; the fillets were sometimes of poor appearance, with the pink brown discoloration that is typical of fish that have been cut before going into rigor, and they would not take on the gloss beloved of fish smokers when subsequently put through a kiln. There were marketing problems too; from the time of her first landing in Grimsby her early catches from north west Atlantic grounds were all handled through Smethurst's of Grimsby, a Unilever subsidiary, and sold at a loss just to develop the market for her products. But attitudes slowly changed, and new outlets were found; Birdseye and Ross were making fish fingers, and the *Fairtry* began making laminated blocks of fillets to supply their processing lines; landings were often at Immingham, a cargo port on the Humber. In March 1957 Salvesen bought a share of a new port cold store, and having decided that *Fairtry* would never be viable as a one-ship venture, planned the building of two sister ships in an attempt to make the experiment show a profit; Harper Gow, Salvesen's managing director, always regarded the whole venture as an experiment, albeit a costly one, even after it was all over.

Fairtry II and *Fairtry III*

Salvesen's had their own cold store in readiness at Grimsby in 1958, the first of what was to become a countrywide chain which continued to provide a public distribution network long after the ships were gone. Simons of Renfrew built both the new ships; *Fairtry II* was completed in March 1959, and *Fairtry III* in February 1960. They were slightly shorter in length than their predecessor, but with greater beam their gross tonnage, 2857, was higher; their cold storage capacity in turn was slightly increased to 680 tons. They had accommodation for 96 people, although 80 was about the number that was usually carried. *Fairtry*

Fairtry II, Salvesen's second factory trawler, built in 1959

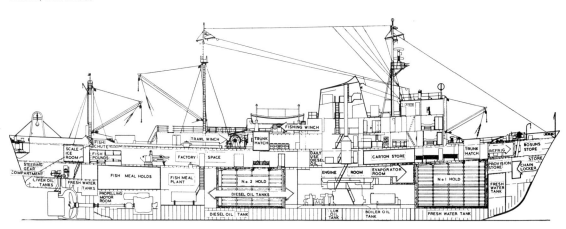

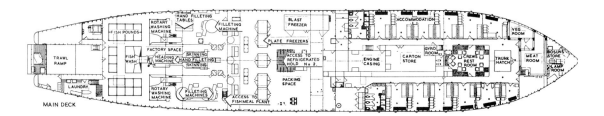

Salvesen advertise their fleet's products, 1960

was diesel driven; her successors were diesel-electric, but were otherwise largely a repeat of a satisfactory model. Sterne's again supplied the refrigeration equipment to freeze 30 tons of fillets a day, and the factory deck was equipped with the latest in filleting and other processing machinery, mainly from the German firm of Baader, the range to choose from having much increased since *Fairtry* was built 5 years earlier.

With three ships in service, continuity of supply was easier, crew reliefs were easier to arrange, and the cost of shore services and other overheads could be shared. From 1959 onwards most if not all fish from the *Fairtry* ships was handled by Ross Group, and Salvesen's had a seat on their board, and in 1960 Salvesen's increased their port cold storage at Grimsby by taking over Northern Cold Stores from Frigoscandia. The whole operation appeared to be prosperous, and the *Fairtry* design was so attractive that the Russians are claimed to have copied it when establishing their own fleet of factory trawlers. But as early as 1962, only 2 years after the third ship entered service, the board of Salvesen's were told that the venture was not doing well, and they seriously considered selling *Fairtry I* at that time. But the *Fairtry* fleet continued to operate until 1968, when it was laid up; the 'experiment' that had lasted 14 years came to an end, and among the many factors that contributed to the closure was the inability of average skippers to achieve high enough catching rates, dwindling stocks on grounds that were yearly becoming more crowded with competitors, and unsatisfactory performance of the diesel electric ships. There were also more general comments about the ships being in advance of their time. What seems fairly certain is that sophisticated processing on the scale attempted on the *Fairtry* ships could not be matched to catching power closely enough to make a profit, in spite of competent management; eastern European countries continue to operate such factory fishing vessels, but since they do not have to sell in a free market and show a return on investment in the same way as in Western Europe, the comparison is invalid.

Fairtry I was sold; *Fairtry II* and *Fairtry III*, both of which were still laid up in Norway in 1970, were also sold eventually, the former for service as a support vessel for submersibles and renamed *Vickers Voyager*, and the latter for scrap.

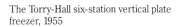

The Torry-Hall six-station vertical plate freezer, 1955

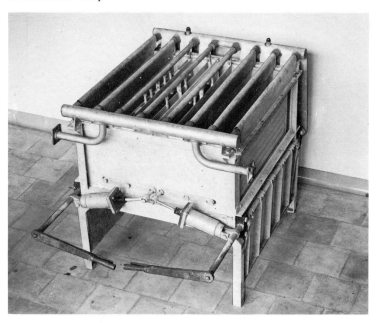

Eighteen months after the launch of *Fairtry* the USSR launched a factory trawler of almost identical dimensions and with the same power and speed. Named *Pushkin*, she went into service in 1955 and had facilities for filleting and freezing her catch, and for making meal and oil from the offal. She was probably the first Russian factory freezer trawler; she was quickly followed by others of the same class, and by other improved classes of vessel that could freeze 30-50 tons a day as against the modest 10-20 tons on *Pushkin*. The rapid and massive expansion of the USSR freezer trawler fleet, and of the fleets of Poland and East Germany, continued throughout the 1960s and 1970s, until by 1975 almost 90 per cent of all Russian frozen fish was being produced at sea.

The Torry-Hall freezer and Northern Wave

After the *Keelby* trials the Torry vertical plate freezer and the associated refrigeration system were much improved as a result of co-operation between Torry and J & E Hall. The new model, known as the Torry-Hall freezer, had six stations instead of four, but was more compact; the pneumatic jacks were replaced by spring jacks, and the movements of refrigeration plates, end plates and bottom doors were all controlled by one large lever. Blocks of fish were discharged from the bottom of the freezer, and could be dropped directly into cold store. Flexible hoses gave way to synthetic rubber O-ring joints in hollow trunnions, and refrigerant was fed through the trunnions to the freezer plates hanging from them. Refrigerant 12 was delivered to the freezer by a special pump driven magnetically through a stainless steel shroud, after much experimentation with pump shaft sealing methods.

While the freezer was being improved plans were made to demonstrate the method on a commercial trawler. In 1953 a layout was designed to fit into that part of a trawler's fishroom that typically remained unfilled, but with a big enough freezing and cold storage capacity to take care of the first 3 days' catch on an average voyage. The remodelled freezer was demonstrated on shore to the industry, the installation plans were presented, and eventually there was agreement on a joint venture. Government, the White Fish Authority and the distant water trawler owners were to share the cost of chartering and converting an existing trawler, and of operating her on a number of experimental voyages. The Grimsby steam trawler *Northern Wave* was acquired in 1955 and fitted out at Hull under the technical supervision of Torry. She left Hull on 31 December 1955, and by the summer of 1956 had completed eight voyages, on each of which some 30 tons of the earliest-caught white fish were frozen and stored at $-30°C$ until landing. The blocks were subsequently stored in shore cold storage, thawed in batches by Torry

Commercial freezing trials on the
Northern Wave in 1956 were a success

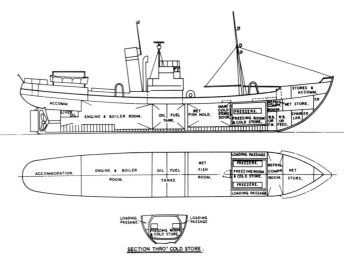

Freezers and cold store occupied the
forward end of the fishroom in the
Northern Wave

staff, and distributed through trade channels to obtain merchants'
reactions to quality and to suitability for processing in a variety of ways.

The *Northern Wave*, built in 1936 and 188 feet in length over all, was of
much the same size and capacity as most of the vessels working in the
UK distant water fleet in post-war years, and if a freezer and cold store
of suitable size could be squeezed into her, then it followed that others
could be converted in a similar manner. On average during the eight
experimental trips, her six freezers, with a modest total capacity of
¼ ton an hour, froze the first 2½ days' catch during the first 6-7 days of
fishing, allowing for up to 3 days' delay in iced buffer storage; the

30 tons of frozen fish landed each time represented about 20 per cent of the total catch, the remaining 120 tons being carried in the conventional manner as wet fish in ice.

In spite of the location of the freezers and cold store in the forward part of the fishroom, and the refrigeration machinery further forward still, right in the bows of the ship, there was virtually no mechanical or electrical trouble at any time, although extraordinarily rough weather was encountered during the early voyages, and the freezers withstood the rigours of rough handling by trawlermen without any difficulty. The *Northern Wave* experiment demonstrated conclusively that freezing at sea of the early part of the catch was technically feasible on a commercial scale, and that sea-frozen white fish were generally acceptable to fish merchants, fishmongers, and the public. But still the trawler owners hesitated; the scale of the experiment had not been big enough to convince them that there was money to be made. The cold chain from ship to shop was admittedly at that time still in its infancy, and the distributors, in spite of their grudging admittance that *Northern Wave* fish had been good, were resistant to change. It took another two or three years to persuade UK trawler owners to have a go, and in the meantime the *Northern Wave* was stripped of her installation and resumed her conventional pattern of fishing.

Hans Pickenpack

West Germany tried freezing the first part of the catch on a side trawler when in November 1957 the 67 metre trawler *Hans Pickenpack* sailed on her maiden voyage. She was equipped to freeze and store 80 tons of fillets before starting to stow the remainder of the catch in ice. She used a Sabroe horizontal plate freezer, with brine at $-35°C$ cooled by a two-stage ammonia plant. The fillets were frozen in 6 kg blocks and stored at $-20°C$. German owners for a time continued to install fillet freezing plant in new tonnage; the *Thunfisch*, which was completed in November 1958, also had a Sabroe horizontal plate freezer that could handle 4 tons of fillets in 24 hours, with ammonia compressors and cold brine, and she was quickly followed by a number of sister ships. By 1960 the German distant water fleet was changing over to stern trawling, and soon afterwards, undecided about the advantages of freezing fillets or whole fish, began installing vertical as well as horizontal plate freezers; the *Kap Farval* of 1963 had three Sabroe vertical plate freezers in addition to a horizontal, which could handle $4\frac{1}{2}$ tons of whole fish and 4 tons of fillets respectively in 24 hours, using R 12.

Persuading the UK industry

By the beginning of 1958 the *Fairtry* had been working for 3 years, and her owners were on the brink of ordering new sister ships, West

Germany had the *Hans Pickenpack* freezing the first part of the catch as fillets, and Torry had demonstrated the *Northern Wave* as a part freezer trawler producing whole fish. Should the UK go for freezing all of the catch, part of the catch, whole fish or fillets?

Torry pursued the arguments it had been making since 1932; a large number of existing distant water trawlers, many of them new, were operating inefficiently, returning to port partly empty because their oldest fish would otherwise become unacceptable. It seemed to make sound economic sense to freeze the first part of the catch, thereby increasing its value, and thus allowing the ship a better opportunity of filling to capacity. After the *Northern Wave* experiment was over, Torry spent a lot of time putting down on paper the projected economics of the part freezer trawler, both for whole fish and for fillets. Since the viability of fillet freezing appeared to depend for success on the use of seaworthy and reliable gutting and filleting machines which were still not on the market, Torry continued to campaign for the freezing of the first part of the catch as whole fish.

A trawler builder was commissioned to make a full design study, using an existing distant water trawler hull form as a basis, with a number of optional propulsion arrangements. Torry had argued for a number of years that a freezer trawler did not necessarily have to be as fast as her wet fishing counterpart, and that the space saved by reduction in fuel and engine space could be more usefully employed in storing fish; it was estimated that a reduction in speed of 1 knot could provide enough cold storage space for another 2 days' fishing. This argument about speed was a tough one for trawler owners to accept; top ships attracted top skippers, and top skippers did not relish the idea of steaming along more slowly than their rivals. For reasons of pride and prestige, the slower ship concept was never given serious consideration, except by one trawling company, and for much the same reasons the idea of converting existing tonnage was also usually discarded; new ventures meant new ships, to maintain the right image. By 1958 two Hull trawler companies were seriously contemplating the freezing of fish at sea, and both were interested only in a new ship built for the purpose. The two companies were J Marr & Son, and Lord Line, a subsidiary of Associated Fisheries; Marr's decided to conduct some shipboard experiments before finally commissioning a new ship, with the result that the Lord Line ship was the first to enter service, and her story is dealt with first.

Lord Nelson

George Reay, Torry's director, and Gordon Eddie, his engineering team leader, managed to persuade Tom Boyd, the managing director of Lord Line, that his next new ship should freeze part of her catch at sea as whole fish. Design work began in 1959 and, partly because Boyd

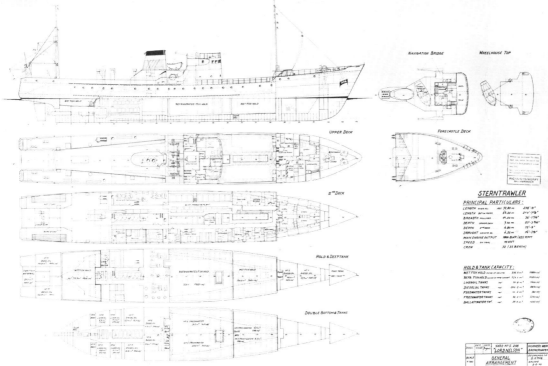

Stern trawler *Lord Nelson*, built in 1961
to freeze part of her catch as blocks of
whole fish.

General layout of the *Lord Nelson*

wanted a stern trawler and Germany by then had some experience in
this field, the order for the ship went to Rickmerswerft in
Bremerhaven. J & E Hall installed the refrigeration plant with Torry-
Hall freezers, and Torry engineering staff stood by the ship as
consultants throughout construction. The *Lord Nelson* was in many

Hauling the cod end up the stern ramp of the *Lord Nelson*

ways a direct successor of the *Northern Wave*; her freezing and cold storage layout followed the same pattern as was demonstrated in 1956, but on a larger scale. Sixteen 6-station freezers produced about 25 tons of frozen, gutted, whole fish a day in 100 pound blocks, which were kept in a cold store that held 180 tons; in addition the wet fishroom held 150 tons of fish in ice. The freezer plates, made from extruded aluminium sections, were an improvement on their predecessors, but the O-ring joints between trunnions and plates were a source of trouble, and eventually had to be replaced by flexible connections in the form of coiled copper tubes. The freezing cycle for 4-inch-thick blocks was 4 hours, including loading and unloading time.

Lord Nelson when she entered service in June 1961 was the first UK commercial trawler to freeze blocks of whole white fish at sea; she was not ideal, but she worked, and many valuable lessons were learnt from her, not only by her owners, but by Torry staff who attended trials and accompanied her on the early voyages. In the event she was the only part freezer trawler to be built, and the only one to use a primary refrigerant in her freezers, but she continued to operate commercially for a number of years, and her owners continued to build more freezer trawlers.

Marbella and *Junella I*

Torry had established the criteria for the freezing and cold storage of newly caught white fish by the early 1950s, and the results of the *Keelby* experiments had been confirmed on a commercial scale on the *Northern Wave*, but J Marr & Son had shore processing and distribution companies linked to their catching activities, and were understandably anxious to confirm that if they froze fish at sea it would

68

meet their own market requirements. To this end they put a single vertical plate freezer on one of their trawlers, the *Marbella*, in 1958, and Torry staff went to sea on her to produce a range of experimental blocks of fish so that the owners could satisfy themselves that the quality was acceptable. The effects of heading, gutting and bleeding on quality were re-examined, and a range of prefreezing delay times was tried on a number of different species. Satisfied with the initial experiments after 6 months, Marr's decided to run a larger experiment. In 1959 they spent about £40 000 on the installation of two 6-station freezers and a small cold store on another of their trawlers, *Junella*, built in 1947, using compressors and machinery built by L Sterne & Co. Sterne's had by this time developed their own design of vertical plate freezer to compete with the Torry-Hall model, having begun with a single station prototype running on R 22 that Torry had tested for them in May 1959. This top loading, end unloading model was in turn based on an earlier prototype developed at Torry for freezing herring.

Junella, sometimes referred to as *Junella I* to avoid confusion with the ship that succeeded her, ran for several trips in 1960 with her pilot scale plant in operation, and was accompanied by both Torry staff and staff from Sterne's. Marr's put the fish through their own distributing company and, as *Junella* fish came to be known and accepted, they decided to use the name for the product and for the new ship they then planned to build.

Junella II

Marr's, like Lord Line, were attracted by the idea of stern trawling, but were not convinced by Torry's arguments in favour of freezing only part of the catch. They opted for a British shipbuilder, Hall Russell of Aberdeen, and ordered a stern trawler that would freeze the entire catch. The old *Junella* was renamed *Farnella* in 1961, and the name was transferred to the new ship, which was completed in July 1962, a year after the *Lord Nelson*. The square midships cross section of a

Junella, built in 1962, was the first UK trawler built to freeze her entire catch in the form of blocks of whole fish

trawler, usually occupied by the ship's engine room, was the best place to build a cold store; Marr's therefore made *Junella* a diesel electric ship, with diesel generators forward, propulsion motor aft and the space in the middle for stowage of the cargo.

The refrigeration system on *Junella* differed from that on *Lord Nelson* in a number of ways. A secondary refrigerant, trichlorethylene, was used in the freezers, as a modern substitute for brine, and was cooled in a heat exchanger by the primary refrigerant, R 22, from the compressors; wasteful and costly leaks of primary refrigerant from the freezers were thus eliminated. The freezer plates were coated with a nonstick plastics film (PTFE), and no hot defrost system was originally fitted; the frozen blocks were released simply by moving the plates apart. There were eleven 12-station freezers with a total capacity of 25 tons of fish a day, although the individual blocks were smaller and lighter than those made on the *Lord Nelson*. Subsequent experiments on Torry's research vessel *Sir William Hardy* showed that defrosting was necessary after each freeze; otherwise wet fish stuck to the cold plates during reloading and prevented the making of a compact block, thus reducing the output of the freezer and the capacity of the cold store. *Junella* was later modified to incorporate a hot defrost, and this increased the capacity of the cold store from 300 to 400 tons with the more compact blocks. The original Sterne freezers were also later replaced by six 20-station Jackstone vertical plate freezers with a total capacity of about 50 tons of fish a day.

Stowing blocks of frozen cod in the cold store of a freezer trawler

When the *Autumn Sun*, a 110 foot trawler from Yarmouth, arrived in that port in January 1960 with 70 tons of frozen herring, she was hailed as the smallest freezer trawler in the world, and as the answer to the problem of handling herring and other small pelagic species caught in areas remote from shore freezing plants; she was certainly small, but she was not the right answer to the problem. Most of her fishroom was converted to cold storage at −20°C, leaving only a small chilled buffer store forward to keep fish that could not be immediately accepted for freezing. Aft of the cold store were eight air blast freezing chambers in which trays of fish could be frozen on shelves. Space was so tight that refrigeration machinery had to be housed in commandeered cabin accommodation under the bridge. Her freezing capacity was designed to be bigger than her catching capacity, as it was intended she should also take fish from other catchers. Since the power supply for the compressors came from the winch generator, the ship could not fish and freeze at the same time, with the result that she operated mainly as a depot factory ship for other catchers, taking fish by transfer in port or in a safe anchorage, a role for which she was not big enough to be economic. She appears to have worked for only a limited period before the system was abandoned.

The third British trawling company to follow Lord Line and Marr into the freezing of white fish at sea was the Ross Group. They began with a conversion in 1963; the conventional steam side trawler *Ross Fighter* was converted to diesel propulsion and equipped to freeze the entire catch as whole fish in vertical plate freezers. This attempt by the

Loading vertical plate freezers with cod in the factory deck of a freezer trawler

company to modernize their existing fleet was not persisted with, and the following year their first new freezer trawler, *Ross Valiant*, went into service. *Ross Fighter* continued to work as a freezer trawler for 3 years, but in 1966 she was converted back to stowage of fish in ice, and the idea of converting existing tonnage was abandoned for good.

St. Finbarr

One British trawling company, Thomas Hamling of Hull, made an attempt to put into practice one of Torry's precepts, the idea of using a slower ship with more cargo space. Their first freezer trawler, the *St. Finbarr*, built in 1964, had a slightly lower turn of speed than her contemporary rivals and, with the engineroom located well forward in the ship, she had a hold capacity equivalent to a vessel almost 30 feet longer. Although the *St. Finbarr* was lost after a fire at sea in 1966, the same general layout was adopted by the company in a further three vessels built in 1968.

Growth in the 1960s, chaos in the 1970s

The general pattern for the UK was established in 1962 with construction of the *Junella*. British trawler owners settled for stern trawlers of 1500-1800 tons gross that could carry 500-650 tons of whole white fish, frozen in blocks in vertical plate freezers that were cooled by a secondary refrigerant. There were variations on the general theme; layout of the ship in particular was amenable to change, and several ways were tried of moving the fish from the stern ramp through reception and preparation areas to freezers and cold storage. As more sophisticated cutting, weighing and mechanical handling equipment became available there were further attempts to move towards the floating factory concept; the Ranger Fishing Company of North Shields built the first of three miniature *Fairtry* ships in 1965, and there were four semifactory ships in Hull by 1968 that could handle part of the catch as fillets with only a modest increase in the size of crew. But in general the UK fleet remained geared to the manufacture of blocks of whole fish for subsequent thawing and processing ashore.

The UK freezer trawler fleet numbered 36 ships by 1970, and grew steadily to a peak of 48 by 1974, but from then on, as nation after nation set out the boundaries of their 200-mile exclusive economic zones and most of the distant water grounds became no-go areas, the purpose for which the freezer trawler was designed disappeared almost overnight. Some ships were sold abroad, others were taken out of fishing altogether, while a few continued to earn a living by combining some white fish trawling with seasonal voyages in home waters to catch and freeze mackerel and other small pelagic fish. By 1982 the number of

Discharging blocks of frozen cod from a vertical plate freezer during fishing

ships was down to 19 and still falling annually; the last new freezer trawler for the UK fleet was built as far back as 1975—and her name?— *Junella*!

Elsewhere in Europe the growth of freezing at sea was rapid throughout the 1960s; West Germany had about 60 ships that could freeze some or all of the catch by the end of the decade, Norway and the Netherlands each had a fleet of about a dozen ships, and Mediterranean countries that could not otherwise reach the Atlantic grounds built freezer trawlers; Greece, Israel and Egypt began fishing distant waters. Spain built the biggest fleet in western Europe, mainly to exploit the rich hake grounds of the south east Atlantic, while the eastern European countries built ships by the hundred; the USSR alone was estimated to have over 500 freezer trawlers and factory ships by the end of the 1960s.

For a time, fishing became an international affair; Japanese freezer trawlers transshipped to carriers in Las Palmas, Russian trawlers worked out of Havana, and Spanish ships were based on Saldanha Bay. Cuban trawlers worked off the west African coast, German trawlers tried South America—the freezer trawler had made distance no object, and inevitably there was strong reaction against the invaders from those whose coastlines were peppered with foreigners. Gunboat diplomacy gave way to conferences on the Law of the Sea. Gradually a

Stern trawler *Coriolanus*, built in 1967 to freeze both whole fish and fillets

Ross Implacable, built in 1968; half the UK freezer trawler fleet was built in the five years 1964-68

The Russian stern freezer trawler *Atlantik*

pattern emerged; those countries which had the capacity to fish their coastal waters to the full did so, and excluded others; those which could not harvest all they had would issue licences or arrange some other means of payment or barter for the privilege of fishing their waters. Countries like Spain and the UK, hitherto heavily dependent on fishing far from base, have largely retreated to home waters, and many of the eastern European ships have taken on the roles of factory ship and carrier for fish caught by others, as in the seasonal transshipment of mackerel off Cornwall and the west of Scotland.

The art of freezing fish at sea, now firmly established commercially for over 20 years, no longer has need of pioneers, but no doubt there will be other firsts, other prototypes, to come as the application of cold to the preservation of fish at sea is adapted to the needs of the industry in the remainder of the twentieth century.

APPENDIX

1835	UK	Perkins and Hogue built the first compression refrigerating machine, using ethyl ether.
1842	UK	Benjamin and Grafton patent: freezing fish in a mixture of ice and salt.
1855	USA	*Flying Cloud* landed and sold fish frozen naturally at sea.
1855	Australia	Harrison patent: ethyl ether compressor.
1859	France	Carré patent: absorption refrigeration machine.
1861	USA	Piper patent: freezing fish in a mixture of ice and salt.
1863	France	Tellier developed methyl ether compressor.
1876	Germany	von Linde built his first ammonia compressor.
1880	UK	*Strathleven* arrived with the first cargo of frozen meat from Australia.
1882	UK	*Dunedin* arrived with the first cargo of frozen meat from New Zealand.
1883	UK	*St. Clement*: first UK trawler with mechanical refrigeration.
1887	UK	J & E Hall made a successful carbon dioxide compressor.
1889	UK	Hesketh and Marcet patent: freezing fish by immersion in mechanically refrigerated brine.
1889	UK	Douglas and Donald patent: freezing fish in a can surrounded by mechanically refrigerated brine.
1892	USA	Fish frozen in cold air, using mechanical refrigeration.
1894	UK	Arrival of frozen salmon by ship from Canada.

75

1898	France	Rouart patent: freezing fish by immersion in glycerol brine.
1905	UK	Kyle patent: precooling fish before freezing.
1910	UK	Henderson patent: precooling fish before brine immersion freezing.
1911	Denmark	Ottesen patent: freezing fish by immersion in a tank of unsaturated brine.
1912	Norway	Dahl developed brine trickle process for freezing fish.
1913	Norway	Bull patent: freezing fish in a can surrounded by refrigerated brine.
1915	Germany	Friedrichs freezing eels in a can of water surrounded by a freezing mixture.
1915	Norway	*Karmoy*: the first fishing vessel to freeze fish at sea by mechanical refrigeration, using an Ottesen brine freezer.
1916	Germany	Plank first showed the importance of quick freezing.
1920	UK	Mann patent: brine freezing tanks for a trawler.
1920	UK	Piqué and Hardy patent: brine drum immersion freezer for fish.
1920	France	de Hervé developed a continuous brine freezer.
1921	USA	Zarotschenzeff patent: Z brine spray system.
1922	UK	Piqué design for a brine drum freezer on a trawler.
1922	USA	Petersen patent: freezing fish in a can surrounded by calcium chloride brine.
1923	USA	Taylor patent: brine spray freezing tunnel.
1924	USA	Birdseye patent: freezing fillets in a can immersed in brine.
1925	USA	Kolbe patent: improved freezing in a can immersed in calcium chloride brine.
1925	France	*Janot*: equipped with Piqué freezer to freeze the catch of accompanying trawlers.
1926	USA	Cooke developed a jacketed contact freezer for fillets.
1926	USA	*Apollo*: depot ship freezing Pacific salmon in an Ottesen freezer.

1926	UK/Norway	*Helder*: mother ship freezing halibut in Ottesen freezer; fish transshipped at sea from accompanying dories.
1927	USA	Birdseye patent: continuous belt contact freezer.
1927	France	*Calgary*: brine freezing fish at sea, as a factory mother ship?
1927	France	*Pen Men*: trawler with brine trickle system to partially freeze her catch.
1928	UK	*Ben Meidie*: fitted experimentally with small brine freezer.
1928	UK	*Arctic Queen*: mother ship, a larger version of *Helder* of 1926.
1929	Canada	*Blue Peter*: depot factory ship freezing fish off Newfoundland.
1929	France	*Sacip*: steam trawler fitted with brine drum immersion freezer, a development of the Piqué original.
1929	USA	Birdseye patent: multiple plate freezer.
1929	UK	*Zazpiakbat*: fitted with Z freezer to take line caught fish not required for salting.
1929	Germany	*Volkswohl*: first trawler built new to freeze at sea, equipped with Ottesen plant.
1929	Italy	*Naiade*: fishing vessel conducting trials of Z system.
1929	UK	*Nunthorpe Hall*: trawler fitted with Sterilex brine freezer; first UK trawler to attempt commercial freezing of her own catch.
1930	France	*Gure Herria*: second French salting vessel fitted with Z freezer.
1930	UK	*Northland*: third and last UK mother ship taking fish for freezing from accompanying trawlers and liners.
1930	Norway	*Alekto*: freezing herring at sea, as a depot ship?
1931	Norway	*Lesseps*: mother ship fitted with Z system for freezing halibut portions.
1931	Norway	*Korsvik*: mother ship taking line caught fish by transfer for freezing.
1931	France	*Jean Hamonet*: trawler converted to freeze her catch at sea, using *Sacip* brine drum freezer.

1932	France	*Marie Hélène*: converted as sister ship to *Jean Hamonet*.
1932	UK	Torry Research Station recommending the freezing of the first part of a trawler's catch.
1934	France	*Fismos*: trawler using the brine trickle system.
1936	France	*Vivagel*: third French trawler fitted with *Sacip* drum freezer.
1936	France	*Pescagel*: fourth and last French trawler fitted with *Sacip* system.
1936	UK	Torry stressing importance of low temperature storage, down to $-30°C$.
1940	Germany	*Hamburg*: depot factory ship equipped with air blast freezer.
1943	Germany	*Weser*: trawler equipped to freeze her catch as fillets in a hybrid air blast-contact freezer.
1944	USA	*Soupfin*: equipped to sharp freeze fillets at sea.
1945	USA	*Betty Jean*: mother ship to freeze shrimp at sea.
1945	USA	*Chirikof*: depot ship equipped to freeze crab and fillets.
1946	USA	*Helen Crawford*: mother ship equipped to freeze at sea.
1946	UK	Burney freezer of the hybrid air blast-contact type.
1947	UK	*Fairfree*: ex-naval ship converted to stern trawling and freezing at sea, using the Burney freezer.
1947	USA	*Deep Sea*: fishing vessel equipped to freeze her catch as fillets.
1947	USA	*Pacific Explorer*: large factory depot ship to freeze transferred catches.
1948	USA	*Albatross III*: research vessel conducting laboratory-scale sharp freezing experiments.
1951	UK	*Keelby*: research vessel conducting pilot-scale trials of the first vertical plate freezer at sea.

1951	France	*Mabrouk*: trawler equipped to freeze part of her catch in a hybrid freezer of the type used on *Fairfree*.
1951	USA	*Delaware*: trawler equipped with pilot-scale brine drum freezer, a successor to the Piqué and *Sacip* types.
1954	UK	*Fairtry*: stern trawler built new to freeze all her catch as fillets, using hybrid and horizontal plate freezing with R 12 refrigerant; first UK freezer trawler to operate commercially.
1955	USSR	*Pushkin*: freezer trawler that closely followed the *Fairtry* design.
1955	UK	Torry-Hall vertical plate freezer developed as a commercial model.
1956	UK	*Northern Wave*: trawler equipped experimentally with vertical plate freezers to demonstrate freezing the first part of the catch under commercial operating conditions.
1957	W Germany	*Hans Pickenpack*: trawler fitted with horizontal plate freezers to freeze the first part of the catch as fillets.
1958	W Germany	*Thunfisch*: second German trawler equipped to freeze fillets; many others followed.
1959	UK	*Marbella*: trawler equipped with single-station vertical plate freezer to produce samples for her owners.
1959	UK	*Fairtry II*: second large factory freezer trawler, similar to *Fairtry*.
1960	UK	*Junella I*: trawler fitted with pilot scale installation of vertical plate freezers to give her owners experience before building a new ship.
1960	UK	*Autumn Sun*: small trawler converted for blast freezing herring.
1960	UK	*Fairtry III*: third and last of UK large factory freezer trawlers.
1961	UK	*Lord Nelson*: Stern trawler built new to freeze part of her catch as whole fish in vertical plate freezers.

1962	UK	*Junella II*: stern trawler built new to freeze all of her catch as whole fish in vertical plate freezers; many others followed.
1963	UK	*Ross Fighter*: trawler converted to freeze all of her catch as whole fish; the only UK attempt to convert existing tonnage.
1964	UK	*St Finbarr*: stern freezer trawler built new to travel more slowly and make more cargo space available.
1975	UK	*Junella III*: last UK freezer trawler built to date (1986), as the fishery for which freezer trawlers were designed began to decline.